3D Printing for Development in the Global South

DOI: 10.1057/9781137365668.0001

Other Palgrave Pivot Titles

David Fitzgerald and David Ryan: **Obama, US Foreign Policy and the Dilemmas of Intervention**

Lars Elleström: **Media Transformation: The Transfer of Media Characteristics Among Media**

Claudio Povolo: **The Novelist and the Archivist: Fiction and History in Alessandro Manzoni's The Betrothed**

Gerbrand Tholen: **The Changing Nature of the Graduate Labour Market: Media, Policy and Political Discourses in the UK**

Aaron Stoller: **Knowing and Learning as Creative Action: A Reexamination of the Epistemological Foundations of Education**

Carl Packman: **Payday Lending: Global Growth of the High-Cost Credit Market**

Lisa Lau and Om Prakash Dwivedi: **Re-Orientalism and Indian Writing in English**

Chapman Rackaway: **Communicating Politics Online**

G. Douglas Atkins: **T.S. Eliot's Christmas Poems: An Essay in Writing-as-Reading and Other "Impossible Unions"**

Marsha Berry and Mark Schleser: **Mobile Media Making in an Age of Smartphones**

Isabel Harbaugh: **Smallholders and the Non-Farm Transition in Latin America**

Daniel A. Wagner (editor): **Learning and Education in Developing Countries: Research and Policy for the Post-2015 UN Development Goals**

Murat Ustaoğlu and Ahmet İncekara: **Islamic Finance Alternatives for Emerging Economies: Empirical Evidence from Turkey**

Laurent Bibard: **Sexuality and Globalization: An Introduction to a Phenomenology of Sexualities**

Thorsten Botz-Bornstein and Noreen Abdullah-Khan: **The Veil in Kuwait: Gender, Fashion, Identity**

Vasilis Kostakis and Michel Bauwens: **Network Society and Future Scenarios for a Collaborative Economy**

Tom Watson (editor): **Eastern European Perspectives on the Development of Public Relations: Other Voices**

Erik Paul: **Australia as US Client State: The Geopolitics of De-Democratization and Insecurity**

Floyd Weatherspoon: **African-American Males and the U.S. Justice System of Marginalization: A National Tragedy**

Mark Axelrod: **No Symbols Where None Intended: Literary Essays from Laclos to Beckett**

Paul M. W. Hackett: **Facet Theory and the Mapping Sentence: Evolving Philosophy, Use and Application**

Irwin Wall: **France Votes: The Election of François Hollande**

David J. Staley: **Brain, Mind and Internet: A Deep History and Future**

Georgiy Voloshin: **The European Union's Normative Power in Central Asia: Promoting Values and Defending Interests**

Shane McCorristine: **William Corder and the Red Barn Murder: Journeys of the Criminal Body**

DOI: 10.1057/9781137365668.0001

palgrave▸pivot

3D Printing for Development in the Global South: The 3D4D Challenge

Thomas Birtchnell

Lecturer, University of Wollongong, Australia

and

William Hoyle

Chief Executive, techfortrade, UK

palgrave macmillan

DOI: 10.1057/9781137365668.0001

First published 2014 by
PALGRAVE MACMILLAN

Palgrave Macmillan in the UK is an imprint of Macmillan Publishers Limited, registered in England, company number 785998, of Houndmills, Basingstoke, Hampshire RG21 6XS.

Palgrave Macmillan in the US is a division of St Martin's Press LLC, 175 Fifth Avenue, New York, NY 10010.

Palgrave Macmillan is the global academic imprint of the above companies and has companies and representatives throughout the world.

ISBN: 978–1–137–36567–5 EPUB
ISBN: 978–1–137–36566–8 PDF
ISBN: 978–1–137–36565–1 Hardback

A catalogue record for this book is available from the British Library.

A catalog record for this book is available from the Library of Congress.

www.palgrave.com/pivot

DOI: 10.1057/9781137365668

Contents

DOI: 10.1057/9781137365668.0001

Foreword

I once asked an eminent colleague, a university professor, what he thought was the most fundamental distinction between things in the Universe. 'The distinction between fermions and bosons', he said. I could see his point: that is the distinction that makes everything from starlight to consciousness work.

But I consider that the most fundamental distinction in the Universe is the one between things like pebbles and things like mice, that is, the distinction between things that exist for a while and things that exist for a while and – during that while – copy themselves. Self-copying gives near-immortality (the DNA in every cell of your body goes back in an unbroken temporal chain nearly four billion years long); it gives intent (every self-copier has evolved either in an unconscious or a conscious desire to reproduce) and exponential growth (when I was born there were three billion people; now there are seven billion). Not all of these properties are without problems, of course, but they are most definitely everything that makes everything interesting.

Self-replication is the most powerful phenomenon on the face of the Earth. Indeed, it made the face of the Earth. And yet, apart from farming and ourselves, we don't make use of it. All of engineering is not like birds making more birds, it is like birds making nests. Nests can be intricate, complicated and beautiful things, but that intricacy, beauty and complexity fade to nothing when compared to the birds themselves. And birds, like all self-replicators, increase in number exponentially given the resources to do so.

Self-replication as a means of engineering production is not a new idea. Queen Christina of Sweden asked René Descartes how a machine might reproduce. And the concept of an artificial self-replicating machine runs through the work of Samuel Butler in the nineteenth century and John von Neumann in the middle of the twentieth century. But it was not until the end of the twentieth century that we had a manufacturing technology so sophisticated that it might stand a chance of replicating itself. That manufacturing technology is 3D printing. That is the reason why, at the beginning of the twenty-first century, I started a project to make a self-replicating 3D printer – the RepRap Project.

Owing not so much to me but much more to a worldwide group of thousands of volunteer engineers, the RepRap Project has become a success and there are tens – possibly hundreds – of thousands of RepRap machines in the world. (Nobody knows exactly how many.) The volunteers volunteered and RepRap has self-replicated itself so successfully[1] because the RepRap Project is free and an open-source. Anyone can download all the designs for a RepRap and make another one, perhaps using a friend's RepRap machine to do so.

* * *

There are big agricultural companies in the world. But none can hold a candle to the likes of Volkswagen and Wal-mart, despite the fact that agriculture is – literally – our only *vital* industry.

The reason for this is that manufacturing and distribution both currently benefit from economies of scale. In contrast, agriculture is *scalable*. Anyone with a little land can grow something.

Growing something from scratch is a lot easier than making a VW Golf from scratch, even though the thing grown is immeasurably more subtle and complicated than a mere car.

However, 3D printing may remove those economies of scale, making manufacturing, like agriculture, scalable and eliminating the need for distribution almost completely.

A *self-replicating* 3D printer starts to make manufacturing much more like farming. And 3D printing means that material goods can effectively be sent down the wires. So 3D printers may start to do for distribution what the Internet has already done for information. (Who reads their news on folded sheets of paper these days?)

Marx pointed out that the rich get richer because they have the capital to invest – directly or indirectly – in manufacturing plants. Whereas the

DOI: 10.1057/9781137365668.0002

poor stay poor because the only thing they have to sell is their labour – a renewable, but not expanding, resource. This is as true for communities and nations as it is for individuals, and Marxist revolutions (inevitably disastrous for all concerned) made the transition from a stable agrarian life to an industrial manufacturing one with much greater inequality.

Yet the most comfortable nations (relatively unequal though their citizens are) are the ones that embraced the Industrial Revolution without a subsequent Marxist revolution. And the poorest and least comfortable nations are the ones that have failed to do that owing, in part, to their not having the capital to invest both in manufacturing and the necessary education that must precede it.

But now, even the poorest people have telephones that give them access to the sum total of human knowledge. And low-cost 3D printers (and the cheapest cost about two phones' worth) are starting to allow anyone anywhere to put a foot on the first rung of the manufacturing ladder.

Adrian Bowyer
RepRapPro Ltd and Bath University, UK

Note

1 In the morning of 29 May 2008 there was only one RepRap. In the afternoon there were two.

DOI: 10.1057/9781137365668.0002

List of Abbreviations

2D	Two dimensional
3D4D	Three-dimensional printing for development
3DP	Three-dimensional printing
BOP	Bottom of the Pyramid
CAD	Computer Aided Design
CBO	community-based organization
CEO	Chief Executive Officer
DIY	do it yourself
FMCG	fast-moving consumer goods
GDP	Gross Domestic Product
GPN	global production network
HDPE	high-density polyethylene
ICT	Information and Communications Technology
ICT4D	information and communication technologies for development
IMF	International Monetary Fund
IUD	intra-uterine device
MDG	Millennium Development Goals
MIC	Middle Income Country
NGO	non-governmental organization
NIC	National Innovation Council
OECD	Organization for Economic Cooperation and Development
PET	polyethylene terephthalate
STL	stereolithography
UN	United Nations
UNIDO	United Nations Industrial Development Organization

DOI: 10.1057/9781137365668.0003

WBG	World Bank Group
WHO	World Health Organization
WOOF	Washington Open Object Fabricators
WTO	World Trade Organization

DOI: 10.1057/9781137365668.0003

palgrave pivot

www.palgrave.com/pivot

Introduction

Abstract: *In this book, Birtchnell and Hoyle ask the provocative question, will 3D printing alleviate poverty to any meaningful extent in the Global South? The authors explore ways 3D printing could offer an alternative to the worldwide production and consumption system and allow objects to be made within circular economies. But queries remain about the ownership of the designs people print, the geopolitics and supply chains of the resources that make up materials for printer feedstock, and the infrastructures printers need to function effectively. Addressing material poverty through 3D printing involves promoting equality of access to the production of objects and this book considers the merits of development at the press of a 3D printer button.*

Birtchnell, Thomas and William Hoyle. *3D Printing for Development in the Global South: The 3D4D Challenge.* Basingstoke: Palgrave Macmillan, 2014.
DOI: 10.1057/9781137365668.0004.

3D printing (3DP) works in the same way as the two-dimensional (2D) printing of pictures or text. A 3D printer is literally a technology that 'prints' out an object in three dimensions from a CAD file. At the low-end of this innovation are 'glue-gun' extrusion or ink-jet powder printers that make objects from heating and depositing plastic connected to a personal computer. At the high-end are laser and electron-beam printers with custom software interfaces, which sinter or melt powdered metals, resins, and other expensive materials in micro-millimetre layers until an object is built.[1]

3DP is being hailed as the harbinger of a third industrial revolution.[2] As with all new technologies, 3DP is the focus of significant palm reading in the media and amongst technology experts about its future mass adoption.[3] A spate of high-profile advocates foresee a rapidly expanding market. For instance, research firm Gartner anticipates that the total number of consumer and enterprise 3DP shipments will grow to over a million by 2017 and users are predicted to spend over US$5.7 billion (£3.43 billion, €4.17 billion) on 3DP between 2014–17.[4] Much of this excitement is not only from the lips of those with a profit to make, but also from academic engineers and scientists, suggesting that 3DP demands careful consideration. These commentators conclude that, because various innovations – the Internet or the mobile phone – have had a notable impact on human development worldwide, so too could 3DP. It is not implausible that 3DP has the capacity to provoke a seismic shift in the way people gain access to objects when compared to the significant development impacts of the mobile phone for communication, the automobile for mobility or the printing press and Internet for education.

Not since the early days of the microprocessor and computer 'chip' has a technology met with the degrees of excitement that 3DP is attracting now. 3DP is a continuation of the microprocessor and chip, allowing computers to convert virtual data into physical matter. This book critically considers the idea that this efficient, precision, decentralized production process – known as 'additive manufacturing' in expert circles – could enable greater access to the means of production for those enduring poverty in the Global South.

The primary difference of 3DP to previous manufacturing technologies is that it produces a tangible physical object without many of the onerous processes observable in mass-manufacturing factories today, which require economies of scale, expensive facilities and safety and expertise

DOI: 10.1057/9781137365668.0004

to make single objects affordably: moulding, milling, cutting, casting, assembly lines, and so on, not to mention transoceanic shipping and logistics. Objects made on 3D printers range from the mundane – toys or novelties – to the critical – motorbike parts or wrenches – and even the exotic – food, organs, and buildings. The process starts with design information in a digital – usually in the Computer Aided Design (CAD) format – file. Computer software enables this digital file to be translated into instructions for the 3D printer. The 3D printer deposits layers of material one on top of another until the object, sometimes supported by scaffolds, is complete.

Is 3DP a technology fix for many of the problems in the worldwide system of production and consumption? Is the attention being paid to 3DP merely a product of hype cycles and snake oil? Is there substance to the notion that 3DP has applications in promoting sustainable communities? In order to examine these questions, this book departs from projections of a 'third industrial revolution'[5] anticipating consequences that lie not in the object-rich Global North, but in the resource-constrained Global South. This book takes an academic, critical stance to the possible social impacts of this innovation in property ownership. Personal property is intrinsically tied up with notions of wealth and poverty and 3DP refashions the processes by which material resources are made into objects and, in turn, made available to people.

3DP in the developing world

In the process of researching this book, we travelled to communities in the Global South and saw firsthand how 3DP has the capacity to make a difference in their lives now. This book also draws on insights from the 3D4D Challenge, which one of the authors, William Hoyle, organized as Chief Executive Officer (CEO) of registered British charity and non-governmental organization (NGO), techfortrade, in 2012.[6] The 3D4D Challenge was the first specifically focused funding intervention that put innovators on their mettle through rewarding the most feasible development projects in the Global South.

A definition of the term 'development' that guides this book is 'the planned attempts to transform the standard of living among the populations of a poorer country or region, generally by outside forces'.[7] In this book, 3DP is thought about as a combination of elements: an 'ecosystem'

DOI: 10.1057/9781137365668.0004

of technologies, materials, designs, and infrastructures, along with sympathetic social practices of sharing, making, and community co- and peer- production. This book considers the interface between 3DP and development and provides depth to understandings of the systemic implications of this particular innovation.

Once coming together in the right configuration, there is evidence that the elements that make up the 3DP ecosystem could meaningfully contribute to standards of living of poorer regions as currently benchmarked in the United Nation's Millennium Development Goals (MDGs), due for a progress report in 2015. These goals are an aspirational 'great leap forward' in efforts to qualitatively meet development in eight key areas of hunger, education, female empowerment, child mortality, maternal health, disease, environmental sustainability, and global partnerships for development. The Challenge and the MDGs share a common ground.

First, Boris Kogan has created low-cost 3D printed greenhouses to provide resource-efficient food production to tackle hunger. Second, the Climate Connected Benefit Society has created solar lamps printed from discarded soda bottles to enhance household light for education. Third, Protoprint has distributed a plastic waste recycler and printer to empower a women's waste-picker union in India. Fourth, Fripp Design and Research address the lack of supply of prostheses for children and adults in developing countries. Fifth, the EN3D Project innovate 3D printed renewable energy systems using design skills from local students and engineers. Sixth, Happy Feet!, in Kenya, 3D print custom footwear to aid sufferers of disfiguring Chigoe flea (jiggers) disease. Seventh, Re-Char WOOF created and deployed the first off-grid 3DP setup, capable of producing tools for small-holder farmers. All of these efforts are global partnerships. In future 3DP will surely interface even more closely with the MDGs.

The idea of 3D printing for development (3D4D)[8] like information and communication technologies for development (ICT4D)[9] and other preceding initiatives is a bottom-up rather than a top-down contribution to these eight goals, one that is more grassroots than governmental. Moreover, 3D4D anticipates development responses that are more balanced and fair through democratization in the production and consumption of everyday objects in those areas currently neglected by globalization.

While certainly not fixing all of these issues in one fell swoop, 3DP's unique features make it a useful tool for resource-constrained communities to draw on and this is demonstrable in the diverse array of projects

DOI: 10.1057/9781137365668.0004

showcased in the 3D4D Challenge. 3DP appears to be something like a 'Swiss army knife' – unlike most other technologies and similar to the Internet in its radical revision of society's norms.

This book locates 3D4D in the canon of scholarly research on ICT4D, 'Development 2.0' and 'post-development'. Responsibilities for meeting development indicators and levelling global inequality appears to have fallen short of aspirations; therefore, this book assesses how 3DP could conceivably change the development game by dislodging access to personal property from the current globalized system of production and consumption and its associated inequalities, oversights, and crises.

Development at the press of a button

'In a developing country like India, 3D printing has huge potential and exciting opportunities which are being explored, like remote printing of products and recycling of household plastic waste for feeding the printers.'[10] In the twenty-first century, regions that harbour manufacturing also harbour poverty and consumers in the Global North are in touch with poverty through the objects they handle and their tags bearing Made in Bangladesh, China, India, Mexico, Pakistan, Thailand, Vietnam, and so on. This book is about tackling poverty, yet what exactly is poverty in the twenty-first century? First, poverty is regionalized according to different dimensions of austerity. In one place austerity is far more burdensome – not having access to food staples say – than in another, where a rise in the cost of student fees attracts public outcry.

Poverty occurs in clusters within distinct regions and some countries are more addled by severe instances than others. As a working definition, poverty is an inability for some people to obtain the staples that other people take for granted across social, regional or even global scales. Clothing, food, tools, property, hygiene aids, transport, and increasingly telecommunications devices (phones, computers, the Internet) all constitute such modern staples. These staples can loosely be understood as fortifying what sociologist Elizabeth Shove calls the three C's: convenience, cleanliness, and comfort. These continue today to be distant for those suffering poverty in the Global South to any standard recognizable in the developed world.[11] Those unable to access the baseline of material objects in the Global North are set outside the systems of worldwide

DOI: 10.1057/9781137365668.0004

production and consumption in a state of informality and so they are prone to unsettling global forces.

The international systems of foreign-aid giving are a complex beast involving all manner of inter-, intra-, and non-governmental organizations working autonomously and in concert. In recent years, the academic focus on development action against poverty has shifted from top-down interventions (in many cases simply the emergency bulk delivery of staples) to bottom-up initiatives that instil resilience in communities in the Global South. In many cases, these aim to motivate 'off-grid' insulation from chaotic global forces. Domestic comfort, convenience, and cleanliness can be achieved in ways that are more sustainable than in on-grid living.[12] This is informative when considering problematic efforts to apply on-grid systems thinking to the Global South. Hence, this book argues that 3DP offers cosmopolitan standards of living that are also off-grid.

The most successful development approach so far is in transfers. Food and medicine are moved to people in need experiencing floods, famine or flight from war. Finances are moved from those places where there is a surplus of wealth to regions in deficit from charity, remittances, venture capital, and investment. Knowledge and technology are moved to foster education and trade. Finally people are permitted to move from developing to developed regions, both permanent and impermanent, as students, retirees, migrants, and refugees.

What goes often unsaid in development discourse on transfers is that most development agendas are tuned to promote a particular way of life modelled on the urban, cosmopolitan standards resplendent in the Global North. This is the developed world ideal: the balancing of global incomes and wealth on a regional or national basis in order that personal incomes, a crucial entry-point into global production networks (GPNs) and consumer markets, can provide the trappings of cosmopolitan life (cosmopolitan literally meaning a 'world citizen'): diverse apparel, personal vehicles, utilities-serviced homes, assistive and networked technologies, convenient domestic goods, appliances, and so on. The most visible sign of a lack of regional development is a lack of access to ownership and use of the assets common in the Global North that give a sense of equity to how the world's resources are utilized by the majority. This is chiefly a material understanding of development and poverty.

DOI: 10.1057/9781137365668.0004

Development and poverty are also understood symbolically as constituting basic access to distinct standards of education, employment, and legality. These facilities notionally support a recognizably cosmopolitan lifestyle as well, ensuring equal access to (generally urbanized) amenities, occupations, infrastructures, and public services. Without sewers, roads, libraries, hospitals, police, courts, parks, and leisure facilities, cosmopolitan life is near impossible to orchestrate to the universally recognized standards of the North.

3DP could potentially satisfy both material and symbolic poverty through transference of both information and objects. Like microfinance, renewable energy or the Internet, there is a considerable degree of speculation about 3DP's potential for growth in the Global South. Yet, 3DP is already being considered as a possible tool for disaster relief in emergency and aid settings where resources are in short supply and social services have broken down.[13] 3DP in the estimation of this book, offers a solution to the disconnected supply chains, collapsed economic markets, and vulnerability of the citizenry characteristic of disaster settings and clusters of poverty in the Global South. This book has little to offer on top-down development. However, development scholars Andy Sumner and Richard Mallett, in scrutinizing the current landscape of development cooperation and its likely future, suggest there are new factors that require a fresh perspective.[14] This leads us to consider the ways in which 3DP could, in fact, meet this demand by offering a bottom-up development option in the Global South.

Disruptive innovations

In 2013, entrepreneur, Kartik Gada, launched the US$100,000 'Prize for Personal Manufacturing', for the first person to invent a truly self-replicating, self-assembling 3D printer. Gada predicts that personal manufacturing offers the chance to reduce poverty, reduce waste, and bring manufacturing to the level 'cottage industries operate in, then the scale of Chinese mass-manufacturing is no longer a requirement to be cost effective'.[15] Gada's vision is to reduce poverty by sparking an innovation economy in the developing world through factories-in-a-box. His idea is essentially that 3DP can uplift a billion people and he believes this could happen within ten years.

DOI: 10.1057/9781137365668.0004

3DP is becoming more affordable and already there are many open source varieties with designs freely available for distribution in online repositories. In this regard, 3DP is a 'disruptive innovation' working at the community and individual levels to allow people to make the items they require at little cost or effort.

3DP integrates well with similar innovations being introduced in the Global South such as the MIT Lab's former director, Nicholas Negroponte's, One Laptop Per Child endeavour.[16] MIT scientist, Neil Gershenfeld, has singled out the Global South in particular as a candidate for the revolutionary use of new rapid manufacturing techniques such as 3DP as a form of intervention via community and grassroots fabrication laboratories known as 'fab labs'.[17] This chimes well with ideas of self-sufficient indigenous efforts to innovation.[18] For instance, leading up to India's Independence in 1947, Mahatma Gandhi imagined India could free itself from British Imperial hegemony through developing individual and collective behaviours and practices using innovative local technologies: the portable spinning wheel making homespun clothing and the solar evaporator producing sea-salt.[19] By drawing attention to the possibility of transition through pragmatic approaches to individual and collective needs, Gandhi successfully invoked support for his campaign for independence and laid the foundations of a future knowledge-economy based on disruptive innovation and out-of-the-box thinking.

This book focuses on a potentially rapid revolution brought on by digital maker technologies in the Global South and the individual and collective uses of 3DP for development. 3DP promises to be a resource for individual and community responses to extant risks in the Global South. The book combines desk research and qualitative data from expert interviews in order to examine the current state of play of 3DP in relation to development and the possible future trajectories of this field of activity. Could 3DP really stimulate grassroots and citizen-led innovation and, ultimately, social change, which complement the Global South's development goals?

The 3DP ecosystem

Many people following the trend of 3DP – and more broadly additive manufacturing – are now contemplating a new form of personal object creation as achievable and even likely in the near future. A 3D printer, like a

DOI: 10.1057/9781137365668.0004

2D paper printer or personal computer, is not just one technology, as from a systems perspective, it is really an ecosystem of autonomous elements in a commensal relationship. There are four core elements in the 3DP ecosystem that would need to become aligned for 3D4D to happen effectively: technologies, designs, feedstock, and infrastructures. Technologies include the printers and components. Designs are the digital information for printing. Feedstocks are the materials made from resources that are the substance of the objects and infrastructures enable the other elements allowing them to move and be made.

This additive process is in sharp contrast to traditional industrial processes of manufacturing where objects are made from reducing large amounts of material in a subtractive fashion, through cutting, moulding, beating and so on. Subtractive manufacturing is usually only cost-effective en masse, but this is not the case with additive manufacturing where only as much as is needed is printed in a distributed rather than centralized manner.

'What then happens to the hundreds of millions of people who will have no employment (not everyone can become "designers, engineers, IT specialists, logistics experts", after all), and who, incidentally, will not have the disposable income to purchase the wonderful products created by digital manufacturing?'[20] A feature in *The Observer* newspaper partly critiques the idea of a possible Utopian social movement stemming from the capacity via 3DP to make objects in the home, office, library or specialist local 'print shop'. Conversely, a review in the *BBC News* describes 3DP as a disruptive innovation with incredible scope for change.[21] *The Economist* predicts a third industrial revolution stemming from this new approach of manufacturing close or near to the consumer.[22] Significantly, *The Economist*, reporting on the winners of the 3D4D Challenge, also cited a 'third-world' dimension to this new manufacturing technique.[23]

Why has so much excitement been brewing in the media about the layering of materials by computers in three dimensions? Partly it is because this is not all that 3DP offers. Reports of the printing of organ structures contribute dramatically to the general awareness of this new technological approach to growing, rather than manufacturing, objects. 3DP captures the imagination because it promises a new paradigm in a world where control over growth (economic and infrastructural) has become a civilizational obsession. Chiefly, the gathering clouds of hyperbole about 3DP imagine it to be the latest step in realizing the far wider

DOI: 10.1057/9781137365668.0004

aspiration of reanimation: of producing living organic matter (organs, skin, limbs, and bones) from scratch according to digital designs. Even more portentously the level of control over processes of making inferred by 3DP – and perhaps even the blended printing of both living and inert matter – offers the chance of a complete overhaul of how production and consumption are exercised and how growth and development are therefore realized.

Much of the speculation about 3DP appears to have evidence behind it. There is something fantastic about the microscopic control computer guidance has over objects within personal space. So too are commentaries on the possibilities of 3DP guided by memories of other less elegant ways of manufacturing that are to all intents and purposes morally questionable. Of factory slavery in the poorer parts of the world; of energetic excesses in transport, consumption and lifestyles; of the dehumanizing assembly line and its discontents; of mass-manufacturing and mass-waste. This moral unrest is informative of how 3DP taps into wider sympathies.

So, in this book, we also consider the morality of 3DP. It is still early days for 3DP due to a number of key hurdles that need to be overcome and these will be detailed in later chapters. However, in this book we argue that these shortcomings need to be re-evaluated in the case of the Global South due to the particular lived realities of these societies. Chiefly, this book examines the idea that 3DP presents alternative intervention opportunities for development in the Global South and addresses the potential of 3DP to meet the expectations of commentators for radical social change and, apropos, development at the press of a button.

The argument of 3D4D

We argue that object ownership in the Global North involves all sorts of complicated systems in manufacturing, assembly and disposal; labour conditions and environmental standards; the geopolitics of resource extraction and supply chains; and the logics and motivations of consumer marketing, branding and corporate profit making. 3DP does not substitute for these entirely. However, many of these systems continue to exclude personal property ownership in pockets of poverty in the Global South. In these areas, 3DP offers a real alternative. Chapters 1–4 will set

DOI: 10.1057/9781137365668.0004

the scene for a survey of 3DP as a rubric of a much wider discourse on the problems of globalization, surveying the various facets of 3DP through analysis of the 3D4D Challenge and setting the stage for the case studies in Chapters 5–6 on 3D4D indicators and forerunners.

If something like 3DP is going to aid the poor then it is going to be like preceding innovations, which can be applied in a resilient fashion within the widespread informality of the developing world. For instance, the mobile phone and its scope for social networking, the automobile and its shortening of distance, or the educational impacts of the printing press. 3DP must also have real impact at a local, community level and be economically and environmentally sustainable without burdensome top-down governance and administration, if it is to contribute to development goals.

Notes

1 Hopkinson et al. (2006) 'Introduction to Rapid Manufacturing'.
2 Lipson (2011) 'This Will Change Everything'.
3 Hopkinson et al. (2006) 'Introduction to Rapid Manufacturing'.
4 Rivera and van der Meulen (2013) 'Gartner Says Worldwide Shipments of 3D Printers to Grow 49 percent in 2013'.
5 Markillie (2012) 'A Third Industrial Revolution'.
6 The brand name of the charity 'techfortrade' is all lower-case.
7 Rogers et al. (2013) *A Dictionary of Human Geography*: 101.
8 William Hoyle coined the term '3D4D' in 2011.
9 Heeks (2009) 'The ICT4D 2.0 Manifesto: Where Next for ICTs and International Development?'.
10 Srivastava (2013) 'Worldwide shipments of 3D printers to grow 49% in 2013: Gartner'.
11 Shove (2003) *Comfort, Cleanliness and Convenience: The Social Organization of Normality.*
12 Vannini and Taggart (2013) 'Domestic Lighting and the Off-Grid Quest for Visual Comfort'.
13 Koebler (2013) 'Is 3D Printing the Future of Disaster Relief?'.
14 Sumner and Mallett (2012) *The Future of Foreign Aid: Development Co-operation and the New Geography of Global Poverty.*
15 Gada (2011) 'Reducing Poverty Through Personal Manufacturing' .
16 Mitra (2010) *Vision India 2020.*
17 Gershenfeld (2007) *Fab: The Coming Revolution on Your Desktop—From Personal Computers to Personal Fabrication.*

DOI: 10.1057/9781137365668.0004

18 Birtchnell (2013c) *Indovation: Innovation in India's Knowledge Economy.*
19 Birtchnell (2012) 'Elites, Elements and Events: Practice Theory and Scale'.
20 Naughton (2012) 'Is 3D printing the key to Utopia?'.
21 Day (2011) 'Will 3D Printing Revolutionise Manufacturing?'.
22 Markillie (2012) 'A Third Industrial Revolution'.
23 *The Economist* (2012a) '3D Printing: A Third-World Dimension'.

DOI: 10.1057/9781137365668.0004

1
The 3D4D Challenge

Abstract: *A worldwide competition pitting innovators in 3D printing against each other in order to produce the most scalable grassroots community action project in the Global South is the principal case study of this book. The efforts of co-author William Hoyle, CEO of techfortrade, are detailed in this chapter on the 3D4D Challenge. Involving a number of different regions the competitors' projects highlight the novelty and breadth of innovation extant in this worldwide initiative to apply what is currently a niche in the Global North to challenges in the Global South.*

Birtchnell, Thomas and William Hoyle. *3D Printing for Development in the Global South: The 3D4D Challenge.* Basingstoke: Palgrave Macmillan, 2014.
DOI: 10.1057/9781137365668.0005.

Beginnings

To date there have been few examples – indicators and forerunners – of 3DP being applied as a development enabler in the Global South. This chapter examines a recent worldwide effort to foster grassroots innovation in concert with overseas intermediaries and supporters in communities of need in Africa, South America, and South Asia. This Challenge makes the case for a serious consideration of 3DP's dissemination and implementation for development. The 3DP ecosystem of technologies, materials, designs, and infrastructures has potential to improve opportunities and services that cover all people; support the policies or context that will improve the lives of the poor; and specifically target the rights, interests, and needs of the poor.[1] This first chapter details how the 3D4D Challenge incentivized a range of groups into thinking through and testing the finer points of 3DP on the ground in the Global South. Moreover, the Challenge sought to foster grassroots movements to work in liaison with intermediaries through knowledge and technology exchanges.

The non-governmental organization (NGO) techfortrade started operating in February 2011. The UK-registered charity's birth was spurred by the 'big idea' that technology innovation could be focused on initiatives that could improve income and livelihood opportunities for people living and working at the bottom or base of the pyramid.[2] More specifically, the original thinking behind the launch of the charity revolved around the initial idea that by applying mobile technology solutions it might be possible to remove the barriers that prevented poor people, often living and working in rural communities, from engaging in trade on a national, regional, or international basis. Much of the early thinking focused on agricultural trade and the challenges of gaining access to reliable market information, making contact with buyers beyond 'the farm gate', and being able to secure deals and receive payment in a secure and reliable way. However, it quickly became clear that in many developing countries, the lack of good quality physical infrastructure, particularly road and rail connections, presents an equally if not greater barrier to trade.

This barrier is not just an impediment to the movement of finished goods, but also to the movement of inputs such as building materials or spare parts. In Sub-Saharan Africa, for example, supply chain management often involves complex transport arrangements that require goods to be moved to and from truck to boat to bicycle. Apart from the

DOI: 10.1057/9781137365668.0005

time taken to transport goods from door to door, the risks of damage or loss are significant. Throughout 2011, techfortrade talked to a few organizations that were working with rural artisans in developing countries, trying to help them market their work and these conversations confirmed the fact that poor physical infrastructure was a significant constraint.

In September 2011, Giles Keating, the Chair of techfortrade's board of trustees and long term observer of 3DP developments for some time, sent techfortrade a copy of an article that had been published in *The Economist* earlier in the year. The article, entitled 'Print Me a Stradivarius' which also linked to another article from the same edition entitled 'The Printed World', included a subheading which read: 'How a new manufacturing technology will change the world'. By the time the article was finished, ideas for the use of 3DP in developing countries were already forming; however, so too were a raft of potential obstacles, not least of which were the costs of the technology and the challenges of erratic power supplies and connectivity issues, particularly problematic if using 3DP was likely to require the movement of large digital files. However, techfortrade was keen to find out more about the potential for this technology in developing markets and, of course, turned to the Internet to see what could be found.

Of course, the answer was precious little. One of the few examples that came up on the use of 3DP for social benefit in a developing world context, was a project called 'Rapid prototypes for Baghdad' (RP4Baghdad) an initiative set up in 2005 by a group of additive manufacturing and rapid prototyping companies to assist surgeons in Iraq who were dealing with complex gunshot and bomb and blast injuries.[3] CT scans that showed damaged bone and highlighted splintering were being sent from the field via the Internet to one of the collaborating partner organizations. Technicians would then adjust the scanned image to remove tissue information to reveal the geometry of the damaged bone so that an accurate 3D-printed replica could be made and shipped back to Iraq within a few days. With a 3D model, rather than a hazy X-ray, surgeons were better able to plan operations, reducing operating times significantly.

With few references available online that supported the hypothesis that 3DP could be a game-changer in developing markets, there was a need for expert advice. Dr Phil Reeves, Managing Director of an Additive Manufacturing consulting company called Econolyst, had worked in the field for over 20 years, having gained a PhD in the mid 1990s.[4] Over a

DOI: 10.1057/9781137365668.0005

couple of lengthy phone calls, Phil patiently explained the current 'state of play' of 3DP and he provided feedback on the three scenarios that techfortrade had been considering. With Phil's help, illustrations of each of the three scenarios were devised to later be used to 'sell' the case for the 3D4D Challenge to the board of techfortrade.

The first scenario was 'Local Component Manufacture'. In many developing world situations, particularly in rural communities, it is often not possible to buy the most simple of components to fix such things as an irrigation pump or a vehicle. In the developed world, lack of immediate access to spare parts is at worst an inconvenience. Across the developing world, it might mean the difference between clean or dirty drinking water or education without electricity. However, a 'telecentre' or community hub, equipped with a 3D printer, a 3D scanner, the Internet, and training, could support localized manufacture of products and components, from simple medical aids or replacement components for agricultural vehicles to parts for generators, pumps or valves. Could it be possible to link together Western design data with developing world manufacturing capability? Could we digitize the reverse engineering process for ever more complex components?

The second scenario was 'Digital Artisans'. Within the developed world, there are over 45,000 3DP systems, ranging from production facilities with 150 machines in one room, down to home-based hobbyist machines.[5] In parallel with the new manufacturing paradigm of factories in people's homes, is a growing community of web-based data sources. Very much in the same way that iTunes allows users to download digital music, web-based portals now allow users to buy and download 3D data files for home fabrication. However, to support potential consumers without their own 3DP machines (of which there are billions), a growing number of online 3DP service centres are also being established, where parts can be selected from catalogues and printed to order for a price.

Most of this consumer-based 3DP activity is focused on the production of toys, jewellery, homeware, and fashion, with the enabling 3D design data originating from Western designers and brand companies, who receive a licence fee for every file downloaded or part printed. In short, the infrastructure exists today for anyone with design skills, access to 3D Computer Aided Design (CAD) software, and the Internet to engage in this global, digital supply chain. So could techfortrade stimulate a community of digital artisans, linked to telecentres in the developing

DOI: 10.1057/9781137365668.0005

world, who are paid for their creativity and design skills, possibly using a fair trade model?

The third scenario was 'Healthcare'. The medical profession exemplifies the coupling of mass production and standardization, with personalization and choice across products such as hip implants and acetabular cups, which are mass produced in different sizes like training shoes. However, many of these products only cater for the needs of a select few, where a perfect set of conditions allows surgery to be undertaken quickly and efficiently. In an equal number of cases, the perfect conditions do not exist. Take for instance trauma, where there is little time for pre-operative planning and less time for bespoke implant design and manufacture.

However, by coupling 3DP with CT and MRI scanning systems, surgeons are now able to make rapid assessment and remediation of patients using personalized implants made from materials such as titanium. To date, 3DP has been used in maxiofacial reconstruction following trauma, hip and knee reconstruction following degenerative bone cancer, and cranioplasty following congenital deformity. Consider the benefits of a 3DP solution in a mobile hospital environment coupled with CT and MRI technologies. Imagine the possibilities in patient care if the orthopaedic factory could be condensed into one small machine unit.

By December 2011, the proposal was ready to be presented to techfortrade's board of trustees. The original proposal focused on three study objectives:

1. To research the three identified scenarios for the use of 3DP for social impact in the Global South, using selected and relevant stakeholder groups in order to explore the challenges and opportunities.
2. To publish the findings of this research in order to inform and educate social impact organizations working in the developing world for whom the use of 3DP might offer quantifiable benefits.
3. To establish an 18-month 'proof of value' trial of the deployment of 3DP technology in a specific scenario in a Global South context. The trial would be used to assess the challenges, opportunities, benefits, and potential sustainability models for deployment of 3DP.

The Board meeting took place on the 22 December 2011. The proposal sparked an animated debate and one of the trustees, Simon Cox, talked

DOI: 10.1057/9781137365668.0005

enthusiastically about the 'X Prize' phenomenon. The Board considered this idea of hosting a competition as well and agreed that techfortrade should find funding for a significant cash prize. As a result, the objectives were amended as follows:

1 To launch an international challenge event with a cash prize being awarded for the best proposal for the use of 3DP for social impact. The prize would be specifically awarded as a direct contribution to taking the proposal through to implementation.
2 To provide useful content to support the challenge which would inform and educate social impact organizations working in the Global South for whom the use of 3DP might offer quantifiable benefits.
3 To stage a high profile event to select the eventual winner which would bring together key technology players, funders, media, and opinion formers from civil society.

By Christmas 2011, the Board had agreed to launch the 3D4D Challenge in 2012.

Spreading the word

techfortrade and other grassroots innovators make use of networks of intermediaries; they are not necessarily large organizations with stocks of resources, personnel, and equipment to muster at the drop of the hat. This leanness is their very strength as they can also take risks that governments, institutions or private corporations cannot. The 3D4D Challenge's worldwide network of intermediaries enabled considerable exposure and leverage, including a US$100,000 prize for the winning entry.

After the Christmas break of 2011, with a fresh graduate joining techfortrade as an intern (thereby doubling its personnel pool), the preparation for the launch of the competition took place. The approach taken was based on a fairly simple set of principles (Figure 1.1).

The 3D4D challenge was launched on 1 May 2012. This gave a small window for techfortrade to plan and prepare a grassroots campaign. Even though the idea of a competition sounded promising, given the earlier lack of success in finding examples of social projects that were using 3DP, the unknown was how much interest the competition might

DOI: 10.1057/9781137365668.0005

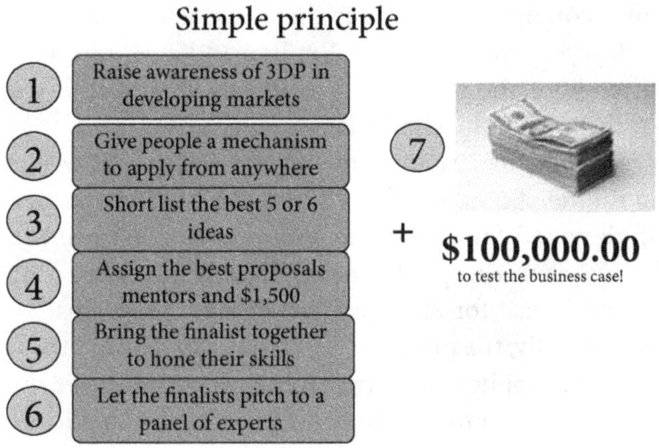

FIGURE 1.1 *Summary of stages in preparation for 3D4D Challenge*

generate. It was clear that the ability to raise the profile of the competition through social networking using digital media would be critical to the success of the initiative.

One of the first tasks was to come up with a name for the challenge. The first effort, 'The 3D Printing in Developing Countries Challenge', seemed too wordy and, most importantly, did not lend itself to a digital campaign. A name that would both function as a logo and website domain while also compatible with social media was needed. Earlier research had included a dive into the 'Twittersphere' to search for contacts. There was already a realization that people tweeting about 3DP tend to use the hashtag #3DPrinting. However, unlike the more established 'technology for development landscape', where the social media convention of using #ict4d or #m4d to distinguish communication about technology and mobile technology in the Global South was an established convention, no such distinction existed for 3DP. Perhaps, slightly tongue-in-cheek, techfortrade adapted the established convention, #3D4D, as its own and it was a short step to settling on the 3D4D Challenge as the overall competition brand.

Hopes to register 3D4D as a domain and Twitter name were dashed by a US-based ultrasound scanning company and techfortrade had to settle for 3D4DChallenge.org. The website was put together very quickly and the initial site was really intended to provide information about the challenge together with informative resources for potential applicants. The resources included a short introductory film[6] recorded by Dr Phil

DOI: 10.1057/9781137365668.0005

Reeves from Econolyst and another short 'endorsement' video featuring Dr Adrian Bowyer, inventor of the RepRap and a definite 3DP 'A-lister'.[7] He agreed to shoot the film after a cold call from techfortrade with a pitch about the idea of the challenge.

One feature that the site did not include was a facility to submit an application online. An opportunity presented itself in another free platform (f6S), designed for challenges and events, which we could simply link to the website. The platform has rapidly morphed into a global online clearing-house for start-ups, accelerator programs, hackathons, and events. Crucially, the platform's functionality allowed the design and deployment of the application form online to communicate directly with applicants and their teams and to provide access for application evaluators to review and rate applications remotely.

The communications strategy really hinged around three strands. The first strand was basically a digital marketing push to mainly manage in-house. The second was a small PR campaign techfortrade outsourced to a small, mainly tech-focused, company called Fourth Day, which was part of the local community happening to share an office with techfortrade. Finally, and perhaps most crucially, a number of innovation workshops were organized in different cities around the world, which would provide an opportunity to engage people on a face-to-face basis.

The digital marketing campaign planning started with some basic online research to identify other NGO's or charities that were using 3DP for social benefit; however, it soon became apparent there was very little and so the search quickly expanded to include academics working in related fields and bloggers and online opinion formers that might help spread the word. Meanwhile, the PR company started to pull together a media list drawn from mainstream, technical and not-for-profit titles and this preparatory work needed to be pretty extensive to cover international as well as domestic media.

The most challenging strand of our communications strategy was always going to be the task of planning the face-to-face workshops. The goal was to spread the net as widely as possible and create and stage events that not only raised peoples' awareness of the possibilities that 3DP might offer, but also encouraged attendees to brainstorm ideas and even form teams that might continue to work on the ideas from the workshop with a view to submitting applications.

techfortrade decided to contact a London-based organization called Social Innovation Camp to investigate how to design the events

DOI: 10.1057/9781137365668.0005

to ensure that the goals were met. Social Innovation Camp designs programmes for private, public, and social-sector organizations that bring people together at creative events to design technical solutions to social challenges.[8] With their experience in the UK and internationally, they would be able to design a workshop framework that could deliver strategic support at the grassroots level. All that was needed was to find the venues and people to participate.

Perhaps the biggest breakthroughs at the planning stages were the sponsors for the 3D4D Challenge. By this stage, Econolyst, Dr Phil Reeves's consulting company, had already agreed to lend their name and contacts to the event. But the hope was to find other industry players that would help us to promote the Challenge and maybe contribute in other ways too, not least of which might be with cash or in kind.

An introduction to MakerBot was facilitated by one of the trustees, John Lazar. A brief Skype call was held with Jenny Lawton, now President of MakerBot. She immediately offered to support the initiative in the form of donated printers and also committed MakerBot to host one of the innovation workshops at their offices in Brooklyn, thus securing the first workshop location. One of her colleagues, John Dimatos was assigned to help organize the event. John's experience as a postgraduate at New York University, working on a programme called 'Design for Unicef', meant that he quickly understood what the 3D4D Challenge was trying to do and he immediately volunteered to co-facilitate the Brooklyn event.[9]

Amazingly, once Brooklyn was secured, the other workshop locations were identified in quick succession. At the suggestion of Social Innovation Camp, techfortrade agreed to host a UK workshop at the Westminster Hub,[10] a central London creative incubator space for social entrepreneurs and innovators that is part of a network of 'Impact Hubs' in more than 60 cities worldwide. Contacts that techfortrade had worked with on other projects were used to secure commitments to host innovation workshops: a hack space in Johannesburg, South Africa, at a venue called House4Hack,[11] in Nairobi, Kenya, at the fab lab based at the University of Nairobi, and in Chennai, India, at the Indian Institute of Technology, Madras.[12] At the very last minute, the plans changed when Chris Worman, a contact in Bucharest, Romania, whose project, ReStart Romania, definitely put him in the online challenges 'expert' category, offered to host an event at the Bucharest Hub,[13] which turned out to be the biggest workshop.

DOI: 10.1057/9781137365668.0005

The web-site went live on the 1 May 2012 and the press release announcing the Challenge was distributed to the media list. The press release was also sent to the mailing lists pulled together from online research and posted on Twitter and Facebook. The release contained details of plans for the innovation workshops and, using the f6S platform, had a facility to enable people to go to the website and enrol for a workshop, stating why they wanted to come and also whether they were already working on a related idea.

The press release and subsequent mailings were widely covered, albeit mainly by trade press such as *Develop 3D*.[14] However, techfortrade did receive a number of requests both to write blog posts about the Challenge and offers from bloggers to write about it. It was one of these offers, from Lauri Poldre who heads the GrabCAD[15] community, which also resulted in GrabCAD coming on board as a sponsor, actively promoting our challenge to their online community of around one million engineers and blogging about the workshops.[16]

techfortrade later discovered that although it had not received much feedback from the mail shots, a number of universities did, in fact, post details of the Challenge internally and it was this that led to the eventual winner. Most importantly, the promotional activity started to drum up interest for attendance in the workshops. Initial hopes were for around 20 people per event to sign up, but this was quickly exceeded at all of the venues. Although the initial momentum was more than hoped for, techfortrade would only know whether or not objectives had been achieved once the applications started arriving.

Building momentum and ideas

In the couple of weeks from launching the competition to the first workshop, the 3D4D Challenge was featured in a number of blogs and news pieces. San Francisco-based, technology non-profit organization, TechSoup, posted the first blog about the 3D4D Challenge in their Netsquared blog.[17] They encouraged people to start thinking about using 3DP for social benefit by pointing out that if a community centre in the middle of a town has public access to computers and the Internet, people come to gather information, learn, and communicate while they develop essential digital skills. So if the community centre had a cutting-edge 3D printer capable of producing end-use component parts – vehicle parts,

DOI: 10.1057/9781137365668.0005

medical implants, telecommunication devices, and even jewellery – local people would then have the ability to create and sell manufactured pieces that were once out of reach or out of range.

MakerBot blogged that 1.1 billion people in this world live every day without a supply of clean water and that 3,900 children die on a daily basis of water-borne diseases. They also encouraged readers to share their ideas for reducing poverty and growing local communities in the developing world. Develop 3D explained that potential ideas for entries could involve using 3D printed models and parts to improve agriculture practices, water supply or filtration processes or energy supplies for rural or impoverished areas. *The Star*, a South African newspaper, expanded on these options, explaining that 3DP has the potential to reduce operating expenses because manufacturing can be done locally. Just like well-established manufacturers, this technology could enable NGOs to create their own working equipment in their offices and other tools required by communities in their working areas.

The Star also reported that the South African Vaal University of Technology was developing a self-help laboratory to be equipped with numerous 3D printers with installed designing software. The objective of the laboratory is to empower students, staff members, and the community to develop their innovative ideas into prototypes.

techfortrade's first innovation workshop took place on 12 May 2012 at the somewhat unusual offices of MakerBot industries on 314 Dean Street, Brooklyn, New York. Arriving in New York to meet John Dimatos and turning into Dean Street, a long queue of people lining up next to a van came into view, which from a distance looked like a soup hand-out for the homeless, but was actually MakerBot's rapidly expanding staff team collecting their free Friday ice cream!

The workshop was held in the production space around the corner from MakerBot's office, the actual space that accommodated the MakerBot assembly line, which bore more resemblance to a cottage industry than a modern day production facility. It was clear that the business was bursting out of its existing facilities and in fact plans were already afoot for a move.

The format for the innovation workshops that had been developed with the team at Social Innovation Camp involved a three-stage approach. Firstly, mixing up the workshop participants into teams to brainstorm and come up with social problems that they might be aware of through their work or through personal experience.

DOI: 10.1057/9781137365668.0005

Secondly, from the list of ideas, each team picked one to work on through stage two which involved thinking about self-sustaining solutions that might leverage 3DP in new ways. The final pass involved teams swapping ideas so that the problems could be looked at through another set of eyes and potentially different approaches considered. Glen Mehn from Social Innovation Camp calls the process itch scratching – 'Someone once said that all good ideas come from having an itch to scratch'.

The group that assembled in Brooklyn was a mix of tech and NGO sector types with a few university students thrown in. The brainstorming produced a long list of 'itches' many of which were echoed in the subsequent events in other cities.

The individual ideas were grouped under categories such as:

- Literacy and sharing
- Water
- Design engineering
- Parts (Recycle, Fixing, Repair)
- Household
- Transportation
- Farming
- Power
- Medical

As the workshops rolled through London and onto Johannesburg, Nairobi, and Bucharest, the themes that had emerged in Brooklyn were expanded with some surprising innovations and ideas being thrown into the mix. It was at the London workshop that techfortrade first met Steve Roberts from Fripp Design, an eventual finalist, who arrived with an example of the soft tissue facial prosthetics that the company had been developing.

At the third workshop at the House4Hack in Centurion, midway between Johannesburg and Pretoria, techfortrade was able to see one of the donated MakerBots in action. Around 25 people from the Johannesburg hacker/tech/NGO communities came together for a lively morning – Schalk Huenis, one of the House4Hack founders, coined the term 'hive', which was discussed quite extensively and refers to the concept of decentralized 'piece work' 3DP to contract.

In Nairobi, thanks to Dr Kamau Gachigi who heads the University of Nairobi Science & Technology Park, which also houses the Nairobi

DOI: 10.1057/9781137365668.0005

fablab, students broke off from the pressure of their exams to engage enthusiastically in the workshop. One of the most unusual and ingenious ideas that the students pitched was the use of 3DP to produce spare parts to fix the many broken condom machines around Nairobi, something which they felt would have direct and positive benefits in terms of reducing the risk of sexually transmitted diseases.

It was also in Nairobi that a group outlined a plan to recycle waste from the Kibera slum to produce 'fair trade' feedstock. The group pitched the ideas stating that a 100 kgs of sorted waste earns a waste picker US$1 yet a kilo of ABS filament sells for US$25.

In all 253 people attended the Innovation workshops, generating hundreds of 'itches' and working enthusiastically on their selected ideas, many of which re-surface in the thinking of the entries that eventually made the way to the 3D4D Challenge application site.

The competition

Once the frenetic activity of the launch at the workshops died down, techfortrade started to turn its mind to the outstanding issue of a venue for the 3D4D Challenge final. For reasons of cost, as well as coverage, techfortrade wanted to embed the final within a large event with a relevant focus. Once again, in early July, Phil Reeves came to the rescue by introducing Kerry Hoggarth, the owner of a London-based events company modestly called, Team Awesome. Kerry's passion for 3DP had driven her to plan a massive and quite revolutionary 3DP show in London in October 2012. The concept was revolutionary because the primary audience was to be the general public and Kerry's intention was to make the show an interactive spectacular that would capture the public's imagination.

From the outset, Kerry was enthusiastic about the idea of including the Challenge final in the show programme. In her email confirming her offer to host the final, Kerry commented, 'our visitors will be enthralled by the 3D4D Challenge and it will also bring a spirit of goodwill and humanity to the show'. On the 18 July 2012, the announcement of the plan to stage the final on the 19 October 2012 was made and detailed arrangements fell into place quickly.

Another of the priorities over the June and July period was to secure a judging panel that would lend weight to the Challenge. techfortrade

DOI: 10.1057/9781137365668.0005

wanted to ensure there was representation from both the world of technology and from experts in the field of development. Using techfortrade's network of contacts and those of the Board of Trustees, a panel of five judges that had credible yet diverse credentials was settled on. Most importantly, they were inspired by the competition and keen to assist. The judging panel consisted of:

- Kai Backman – a co-founder and CEO of Tinkercad, a leading web-based CAD platform, which in 2012 claimed to hold the largest public repository of solid models on the planet.
- Rupert Goodwins – then editor of *ZDNet UK*, a technology news and reviews web site, and previously technical editor of titles such as *MacUser*, *PC Magazine*, and *IT Week*.
- Steve Haines – techfortrade trustee and Mobilization Director for global campaigns at Save the Children International. Steve was previously country head of the Africa Governance Initiative in Rwanda, where he served as an advisor to the Government of Rwanda on building capacity for economic development.
- Mariéme Jamme – a Senegalese-born and London-based global development activist and social entrepreneur, co-founder of Africa Gathering and a regular contributor to *The Guardian* newspaper's Global Development section.[18]
- Simon Trace – Chief Executive of Practical Action, with 20 years' experience at the NGO Water Aid, firstly on soil and water management, drinking water and sanitation in South Asia, and then on a series of posts at its headquarters, including six years as International Operations Director.[19]

A venue for the final and a panel of judges were ready. All that was needed now was the finalists.

The deadline for applications was 31 July 2012, but as the deadline day approached, it was clear from monitoring the online submissions that quite a few of the applications were still a work in progress. techfortrade took the decision to extend the deadline by a week to the 6 August 2012 to give as many applicants as possible the chance of crossing the line.

At the cut-off date, 70 submissions had been gathered and John Lazar, a techfortrade trustee, and Phil Reeves would assist the team in the task of evaluating the applications in order to produce a final shortlist. From the outset, the winning idea would be the one that stood out because it achieved three things. First, it compellingly and measurably addressed

DOI: 10.1057/9781137365668.0005

a significant social need. Second, it demonstrated real technical innovation, in other words, it brought together leading edge thinking across mobile, web, manufacturing, and solar technologies. Finally, it clearly articulated a business model that underpinned the idea of sustainability, not simply relying on continual donor funding.

The geographical diversity of the applicants was enormous, with submissions from countries such as Brazil, Denmark, India, Kenya, South Africa, and the USA. There was also a massive range of experience evident amongst the contestants, from experienced designers such as Ronen Kadushin, a Berlin-based designer and inventor of the 3D printable 'Bearina' intra-uterine device (IUD), to Weke Ziggy, a young Nairobi student with a passion for preserving African indigenous musical instruments.[20]

However, evaluating the applications revealed a number of consistent themes:

- The creation of 3D design marketplaces linking local designers in either developed or developing countries with their customers.
- The development of digital design libraries, using crowd sourcing to obtain design support and facilitating production on a royalty free basis for a range of goods needed by communities.
- Equipment for recycling plastic waste to produce filament which might then be used for local printing.
- Micro-Enterprise networks of 3DP kiosks.
- Product designs to address specific problems such as a low cost printable water filter that could be attached to the neck of a soda bottle or a process for printing customized glasses for wearers with asymmetrical faces, enabling correct alignment of corrective lenses.
- New approaches to construction using a catalogue of printable interlocking bricks that remove the need for other materials to hold the construction together.
- Education facilities to equip communities from childhood with the skills required to use 3DP technology for the benefit of their community.

In the end, seven finalists were selected, rather than the originally planned five. It was a tough choice, made tougher by the fact that there was a relatively small amount of information (one or two sides of A4) about each project. What follows is a short summary of each finalist's project.

DOI: 10.1057/9781137365668.0005

Boris Kogan (USA/Israel)

Boris's project set out to address the demand for food through the production of a small-scale, easy to manufacture and assemble robotic greenhouse using an Arduino computer board (locally fabricated using either 3DP or milling) to maintain an optimal internal environment and water/feed the plants growing inside as necessary. The greenhouse would be built as a geodesic dome using a combination of widely available materials for struts and 3D-printed connecting elements. Depending on the sensors' inputs, the Arduino board would control various elements including 3D-printed vents, solar shades, cooling fans, swamp coolers, and heaters to provide an inexpensive and robust controlled environment for organic production in a variety of climates.

Boris's application argued that the benefit of his design was that it would easily lend itself to hydroponic and aeroponic production, as well as a combined aquaponic/hydroponic system which would grow fish and vegetables in a 'synergetic' manner. The project was designed to scale, from a backyard or front lawn to a larger, farm-sized operation and Boris claimed that the yield of an initial, small investment in a single dome could be used to pay for more and/or larger domes.

Climate Connected Benefit Society (UK)

Edmund Bell-King and Cornell Jackson from the Climate Connected Benefit Society proposed that their ColaLight project would benefit rural communities in developing nations unable to gain access to on-grid electricity, by providing solar lamps produced through the reuse of Coke bottles replacing the use of kerosene. Inside the ColaLight soda bottle would be a rechargeable battery, battery attachment, and an attachment for a printed circuit board which would contain battery connectors, LEDs, a charge controller, and switch connection to allow the lamp to be turned on and off. The bottle cap would contain a charger attachment with a cable that protrudes out of the bottle cap and connects the battery with an external solar panel. 3D printers would be used to produce the bottle cap, charger attachment, battery attachment, and PCB attachment locally to allow lower cost of production, delivery, and reduce supply chain emissions. The team were proposing a trial of their initiative in Tamil Nadu, South East India, in partnership with the University of Pondicherry and they estimated that the project could become sustainable within 18 months, conditional on the sale, through various channels, of 100,000 units.

Fripp Design and Research (UK)

Tom Fripp and Steve Roberts of Fripp Design had plans to address the lack of supply of prostheses for people in developing countries who require them due to disease, accident or deliberate disfigurement. Their project had started back in 2007 when the University of Sheffield approached Fripp Design and Research about the possibility of using 3DP to rapid manufacture soft tissue prostheses. Fripp undertook a feasibility project and were able to manufacture some prototype parts on a 3D printer. The university and Fripp then secured funding from the Wellcome Trust for the research necessary to find out if it could be possible to develop a method to rapid manufacture soft tissue prostheses using the latest industrial design methodologies. Their application to the Challenge was based on their desire to undertake further work to enable the developing world to manufacture soft tissue prostheses for themselves, which would require additional development in the area of 3D data capture to eliminate the need for an expensive 3D scanner. The Fripp team believed that once the additional technical challenges had been overcome, it would be possible to set up centres in the developing world in weeks (which would include the purchase, installation and the training in how to use the technology). Whereas a UK manufactured prosthesis costs over £1000 to make, their aim was to make it feasible to produce an equivalent product for under £10 in a developing country, thus potentially enabling tens of thousands of people to have their lives transformed.

J. F. Brandon – The EN3D Project (Canada)

J. F.'s EN3D Project intended to use 3DP to improve and expand access to renewable energy sources in Bolivia and then the rest of the world. The initiative, centred around a local fab lab would carry out R&D into 3D Printable Renewable Energy systems, using crowd sourced design skills from platforms such as GrabCAD and prototyped by local students and engineers. The necessary parts for semi-custom installations would then be printed by a team travelling from village to village as sales were made. J. F. had developed a simple solar tracker that he claimed was efficient, cheaper, and easy to make than existing models, using a specially designed 'groove' to guide a strut that tilts the panel to face the sun perfectly. The groove is angled to match the sun's path in a specific geographic area. Using a 3D printer, you would have a custom solar tracker with one motor and a timer.

The Pai family: Just 3D Printing/Protoprint (India)

The Pai family, Suchismita, Jayant, and son Sidhant, submitted their idea for 'Just 3D Printing'; a social business to be based in Pune, India, that would source recycled plastics procured from local waste-pickers. Their initiative involved the development of three machines. First, a 'Flakerbot' to be used to transform the waste plastic products into 'flakes' or 'grounds' using low-tech mechanical means while doubling the price of the product and fetching the waste-pickers more money. Second, building a 'Refilbot' that would convert these grounds/flakes into filament that can be used for 3DP. The filament thus produced could either be sold to 'Just 3D Printing' or to others requiring filaments. The final step involved designing and building affordable do it yourself (DIY) 3D printing machines that could be replicated in developing countries like India and installed in kiosks at popular locations all across the country. The kiosks would be geared towards young entrepreneurs and students and would be fairly priced and subsidized so as to allow rapid and low cost prototyping.

Roy Ombatti – Happy Feet! (Kenya)

Roy's 3D4D project concept involved the manufacture of shoes from recycled plastic for the jigger-infested population of his country, Kenya. The jigger is a miniature sand flea measuring about 1 mm and found in unhygienic environments in tropical and sub-tropical regions. Once on the skin, the female jigger burrows itself about 5 mm into the host's flesh under the nails where it can lay over 100 eggs or more in 30 days. An itchy sensation is felt which, if scratched, ruptures the egg sac spreading the infestation. The jiggers and the hatchlings live off the host's flesh and blood, crippling any part of the body that becomes jigger infested and causing septic sores. The open sores leave the body prone to secondary infections such as tetanus occasionally leading to HIV transmission through sharing of needles used to remove the jiggers. Roy, a Nairobi University student, planned to work with the Ahadi Kenya Trust, the sole organization in Kenya fighting this ailment, to print specially adapted orthopaedic shoes made from recycled polyethylene terephthalate (PET) waste.

DOI: 10.1057/9781137365668.0005

Re-Char WOOF (Washington Open Object Fabricators) USA

A last minute joining of forces, encouraged by the evaluating team, led to two groups with similar ideas to be selected as finalists. The logic behind the decision was based upon the observation that the combination of the two teams strengthened the project as WOOF had plastic re-cycling experience and Re-Char had local deployment experience in Kenya. The newly combined team proposed to create and deploy the first off-grid 3DP setup, capable of starting with waste plastic bags and producing tools for small-holder farmers to reduce their carbon footprint while improving their yield. The system would require no connection to the electrical grid and be capable of producing a majority of the parts necessary to duplicate itself.

The proposed benefits of the initiative included the monetization of waste, in the form of plastic bags which litter the streets of Nairobi's biggest slum, Kibera and many other slums. Using waste bags as a feedstock means it would be possible to pay workers to clean up this pollution. Local deployment of flexible 3D printers could mean that those in need might learn how to use 3DP while deploying and creating their own sustainable products.

The final and the fallout

On 12 September 2012, techfortrade announced the names of the finalists. It also announced that each team would receive US$1000 and access to expert mentors in order to develop their projects in the run up to the competition final. For the first time, techfortrade revealed that the finalists would be recording videos of their pitches for viewing at the show and that the final itself would be held live at the show and also simultaneously video streamed for an online audience. In only five weeks' time the finalists would arrive in London for the live final and a chance to win US$100,000.

All of the judges agreed to write about the 3D4D Challenge and post their blogs in the weeks leading up to the final, generating more interest in the event and its goals. These blog posts are included to illustrate the reasons why techfortrade has been so energized by the 3D4D concept from the start.

The finalists began to arrive in London on 17 October 2012. Caroline Fox, a freelance event organizer who joined techfortrade as the organizational workload mounted in the couple of weeks leading up to the final, arranged for a meeting of the group for dinner that evening in order to brief them on plans for the forthcoming three days. This was the first time that the finalists had met together and with techfortrade face-to-face. One immediate observation that was obvious was that the various 3D4D groups were eager to discuss each other's projects and extend offers of help or advice, regardless of the competitive nature of the Challenge.

techfortrade had set aside the 18 October 2012 for the finalists to rehearse their pitches and to record short 'talking heads' videos, each describing their projects. With both a live audience and live streaming of the content and a judging panel to convince, techfortrade was keen to ensure that all of the finalists were able to deliver a polished presentation and had no knowledge of the prior experience that each of the finalists might have of this type of situation. On the recommendation of one of the board members, techfortrade decided to hire a 'pitching coach' to work with each finalist and give them a chance to perfect their performance on the day. Annette Kramer is an American Performance Coach, based in London whose work has helped entrepreneurs and business heads secure funding, from the boardroom to the BBC's *Dragons Den* television show.

Annette worked with each team through the day and in the spirit of collaboration, most of the teams stayed around to watch and offer encouragement and support. As each finalist finished their pitch, they were immediately whisked away to film their video summaries, which were to be edited overnight to create one summary of the rehearsal day which would be shown at the London 3D Print Show in a small 3D4D Challenge cinema on the Show floor. The day was pretty gruelling, especially for those that had travelled long haul only the previous day. Although the techfortrade team was concerned that Annette's tough yet constructively critical approach might undermine the confidence of the presenters, she clearly knew her craft as all of the finalists felt that the coaching really helped to hone the delivery of their pitches.

The 3D Print Show was being staged at The Brewery in London. The site of a former working brewery which operated from the 1750s until the 1970s, the building is now used as an exhibition, conference and events venue. Day One of the 3D Print Show was due to start at around midday on the 19 October with the venue open only to press, trade and VIP guests. Some of the finalists had been asked to carry out interviews

DOI: 10.1057/9781137365668.0005

with TV or press over the course of the afternoon, but for others it was just a case of waiting with jangling nerves.

The final itself was scheduled to take place at 5:00 P.M. Bre Pettis, the co-founder and CEO of MakerBot, one of the sponsors, had generously agreed to allow techfortrade to stage the final in the 'MakerBot Room', a large space off the main exhibition floor, where MakerBot were announcing the launch of their Replicator II. In order to set up the room for both an audience and live video streaming, the room needed to close for a quick re-fit before the final could start.

At just after 5:00 P.M., following Bre's welcome and a quick introduction, the final was underway. Each finalist had just ten minutes to present, and then fielded questions from the judging panel. The audience made up mainly of exhibitors and journalists were clearly impressed with the polished performances and the whole event, which actually lasted for 90 minutes, went by in a flash. While the finalists and guests adjourned to the drinks reception that preceded the VIP evening event, the judges were dispatched to a private room to deliberate. At around 7:00 P.M., the judges made their decision and the names of the first, second, and third prize winners were placed into sealed envelopes. As the VIP evening event got underway and models walked the ramp in 3D-printed shoes while a band played 3D-printed instruments, our finalists waited nervously behind the scenes for the finale of the event, the announcement of the 3D4D Challenge winners.

At 8:15 P.M. as the evening drew to a close, techfortrade's Chair of Trustees, Giles Keating, got up on stage and opened the first envelope. In third place came J. F. Brandon's EN3D project and in second place were Tom Fripp and Steve Roberts from Fripp Design. But the overall prize winners and recipients of the US$100,000 Challenge prize were the team from the University of Washington, Bethany Weeks, Brandon Bowman, while Matt Rogge – Team WOOF.

Our press release announcing the winners went out on the 20 October, 2012 and again on the 22 October 2012. Although various broadcasters and journalists had expressed interest in the final, techfortrade had no idea how much coverage would make it to publication and so were delighted when interviews with Boris Kogan and Roy Ombatti made it to a montage about the show in the technology section of the *BBC News* website.[21] Not only that, the UK's *Independent* newspaper ran an article entitled 'What's the big deal with 3D printing?'[22] The article read:

> Thankfully, a highlight of the first day of the three-day show was the MakerBot-sponsored 3D4D challenge, in which UK charity techfortrade challenged designers and engineers to use 3D printing to help the developing world' and 'the 3D4D presentations proved that, given its lack of limitations and relatively low costs, the technology "has the ability to empower people and change the world", according to Pettis. With researchers already recreating human tissue and hoping to work out ways to "print" circuitry (so you could print off an iPod, say), you sense he's not far wrong.

Perhaps most impressively, on 3 November 2012, *The Economist* ran an optimistically slanted full-page print and online article, featuring our winners, entitled 'A Third World Dimension' with a sub-headline that read: 'A new manufacturing technique could help poor countries as well as rich ones'.[23] It was a pleasing symmetry given where techfortrade had started the journey back in September of the previous year.

Notes

1 In much the same fashion as information and communication technologies for development (ICT4D). See Heeks (2009) 'The ICT4D 2.0 Manifesto: Where Next for ICTs and International Development?': 3.
2 The phrase 'bottom of the pyramid' was first used by U.S. President Franklin D. Roosevelt on 7 April, 1932 during a radio broadcast in which he spoke about *The Forgotten Man.*
3 Dvorak (2006) 'Rapid Prototypes for Baghdad, One Year Later'.
4 http://www.econolyst.co.uk.
5 Estimate for 2010.
6 Econolyst (2012) 'The 3D4D Challenge Explained'.
7 Wilcox (2012) 'RepRap and the 3D4D Challenge'.
8 http://sicamp.org.
9 http://unicefstories.org/partners/academia/.
10 http://westminster.impacthub.net.
11 http://www.house4hack.co.za.
12 http://www.iitm.ac.in/about.
13 http://www.impacthub.ro.
14 http://www.develop3d.com.
15 http://grabcad.com.
16 http://blog.grabcad.com/2012/05/the-3d4d-challenge/.
17 http://www.techsoup.org/.
18 http://www.africagathering.org/.

DOI: 10.1057/9781137365668.0005

19 Practical Action was formerly the Intermediate Technology Development Group, founded by the influential social economist, E. F. Schumacher, in 1965.
20 Kadushin (2013) 'Bearina IUD Concept'.
21 Shaw (2012) 'Bras and Guitars: What People Today are 3D Printing'.
22 Dean (2012) 'What's the Big Deal with 3D Printing?'.
23 *The Economist* (2012b) 'A Third-World Dimension'.

DOI: 10.1057/9781137365668.0005

2
What is 3D Printing?

Abstract: *3D printing is a novel approach to producing objects in a manner reminiscent of other, now ubiquitous, digital technologies: the desktop scanner and the paper printer. Birtchnell and Hoyle determine that 3D printing offers an alternative to the worldwide consumption and production system in some key areas that impact upon material poverty. Establishing that there is momentous potential here, Birtchnell and Hoyle review the debate about 3D printing's place in a possible next industrial revolution. 3D printing is in fact a diverse range of elements including printer technologies, digital designs, materials refined from resources and infrastructures providing energy and logistics. The authors highlight that it will be necessary to get 3D printing 'just right' in order for it to impact upon development goals.*

Birtchnell, Thomas and William Hoyle. *3D Printing for Development in the Global South: The 3D4D Challenge.* Basingstoke: Palgrave Macmillan, 2014.
DOI: 10.1057/9781137365668.0006.

 DOI: 10.1057/9781137365668.0006

Wealth without money

At its heart, 3DP offers personal control over the production of objects (classically known as 'the means of production'), a phenomenon that declined dramatically throughout the world in the twentieth century, particularly in the Global North. While homelessness, illiteracy, disease, drug abuse, and other features of social inequality continue to haunt the fringes of the developed world, *material poverty* is now greatly reduced in high-income countries due to the surplus of affordable and disposable objects available to consumers at paltry cost.

However this book shows that material poverty is far from extinct in the world and continues unabated in pockets of the Global South. This chapter considers the appeal of 3DP as a response to enduring material poverty in regards to the failures of the most efficient method of manufacturing in history: the factory assembly line. The success of Fordist production was due to two parallel system innovations in the twentieth century. First, 'containerization': the global transportation of standardized objects through long supply chains, cheap energy and complex computerized logistics, inventory and distribution systems. Second, 'financialization': interconnected markets and currencies powered by globetrotting traders, investors, shareholders, and financiers, which allow some to profit to an overwhelming degree from the wholesale conversion of resources into objects. These forces have had a profound consequence in the levels of personal property ownership, energy-use, and material consumption in the Global North; however, they have not had a universal effect across the world and in fact have exacerbated material poverty in some regions due to the inequalities they promote and profit from: labour exploitation, environmental degradation, and the privatization of common land.

The inventor of open source 3D printer the RepRap, Dr Adrian Bowyer, proposes a novel idea that captures this book's core concern: wealth without money. Bowyer points to the ability of 3DP to create material wealth, in terms of the ownership of goods, without the need to sell labour in order to participate in the Global North's consumer markets. This chapter shows that 3DP is in a unique position to benefit clusters in the Global South where personal wealth is low. These clusters remain currently out of reach of the factory, the stock market or the supply chain and a keystone idea of this book is that 3DP offers a compelling response to enduring material poverty through enabling an alternative for those outside of

DOI: 10.1057/9781137365668.0006

these systems. So how did 3DP arise as a prime candidate for development action from its roots as a niche approach to producing objects?

3D printing evangelism

'These achievements of 3D printing are not in question. But the application of 3D printing will not be helped by its aficionados getting evangelical about it'.[1] Regardless of the hyperbole, many respectable sources now claim that 3DP will 'change the world'.[2] Bringing the consumer closer to production promises a new intimacy to these commentators that has been lacking ever since the early twenty-first century introduction of mass-machine labour to craft industry. Hence, the now frequent references to peer- and co-production through 3DP. Much like the early 1980s computers, 3DP has provoked journalists to predict imminent social change. As well, there are similar movements amongst the major companies in 3DP towards heroic innovators, mergers, and acquisitions – 3D Systems's takeover of Stratasys is a notable milestone.[3]

At the low end of 3DP are consumer units now available from high street and online retailers with minimal set-up requirements targeted for home usage. The main uses at the low end of the market involve the creation of custom (also termed 'bespoke') objects and parts of an experimental nature. There are a number of limitations in the 3DP ecosystem and these particularly apply to the low end, consumer home printers: difficult-to-use design software; market-standard object finishes and qualities; expensive material feedstock; and limited colours, materials, and mixtures, as well as limits to build-tray sizes. These do not appear to be deterring commentators' enthusiastic reports of growing numbers of early adopters, despite some industry experts noting a similarity to the craze for bread machines in the 1990s – now far from a ubiquitous technology – with many similar functional elements and challenges.

While 3DP evangelism certainly contains a significant degree of hype, there are common features in this arena that point to some or all of the elements within the 3DP ecosystem being useful for development in the Global South. Despite there being an almost incomparable degree of difference between the low end and high end, a unifying trend across both is the idea that manufacturing and, more generally, systems of production and consumption could be positively warped or completely circumvented in the Global South. Central here are the early adopters

DOI: 10.1057/9781137365668.0006

and what can be termed 'maker movements' who, in experimenting with 3DP, are driving the innovation forward. Certainly many of the uses that these movements put to 3DP are hardly conducive to development – the production of novelties (cake decorations or lifelike dolls) being a case in point. Yet, even a cursory survey of online and free-to-use design repositories reveals some objects compatible with development goals. More deeply, the severance from financialization and containerization 3DP affords offers a crucial development angle to this innovation.

The next Spinning Jenny

'Not since the Spinning Jenny ushered in the Industrial Revolution has a piece of machinery been talked about with such hype.'[4] At its heart, 3DP – like the incremental mechanization of industry in the nineteenth century such as the Spinning Jenny (wordplay on 'engine') loom – promises considerable social reform. 3DP is not simply a single product offering a cascade of previously unheard of applications to the masses. Recent excitement about 3DP has its roots in key elements within the 3DP ecosystem, which have become financially accessible for tinkerers, students, researchers, and hobbyists. What is distinguishable in 3DP is the convergence of a 'low end' and a 'high end'. Currently, the low end is oriented towards consumers and prototyping while the high end towards engineers and scientists. The latter range of printers remains the preserve of a small number of intensive users while the former is now becoming increasingly accessible. 3DP, however, does have precursors in history, as noted by Sociologist Anthony Giddens:

> I think we are in the middle of one of the most momentous transitions in production that has ever been witnessed in human history. It is so early that most people I think don't have a real sense of its transformative character ... it's hard to exaggerate what a transition this could be. The early version of it is 3D printers and everyone will know that 3D printers can already produce an amazing diversity of objects from engineering parts to dental crowns and many, many other things. But 3D printing is like Arkwright's Spinning Jenny. It is just the early edge of a world in which increasingly computers will make the world.[5]

Giddens flags 3DP as an example of a technology that brings the virtual into the physical world, in the process becoming a method for re-industrialization and re-shoring in the Global North. 3DP is then 'likely to

DOI: 10.1057/9781137365668.0006

produce a revolution in manufacturing and the point of it is you can bring it back locally, you don't need to import from China or import across the rest of the world, you can do it locally through your computer but within a global network'.[6]

Similar to Giddens, there are also a number of third sector organizations and initiatives targeting 3DP as an intervention capable of bringing about a significant transition in how development goals are met from the grassroots level up, principally in the Global South, which has the lion's share of poverty.

Nevertheless, as the UK-based consultancy and research firm Econolyst emphasizes, consumer low end 3DP for the most part still 'sucks'.[7] This is due to the limitations of the printers, materials, and designs and is not surprising – the development of the personal computer in the 1980s was also incremental. Regardless of this scepticism at the low end and lack of scale at the high end, enthusiasts claim that 3DP has the potential to be 'bigger than the Internet' once a middle ground materializes: Econolyst notes that industry estimates point to 60,000 home and 30,000 commercial users globally in 2012. This adoption points to a healthy 250 per cent annual growth rate in comparison to the Internet's 46 per cent annual growth in 2012. If sustained this would lead 3DP to be on par with the Internet in 2025.

University of Nottingham Engineer, Chris Tuck, makes it clear: at the moment 3DP is far from quick due to the very nature of its processes.[8] Layers of filament, powder or liquid are energetically deposited with a mechanism to bond the different components together, the latter taking time to take place with enough strength to be used. As Tuck notes, although speed has increased in 3DP over the last decade, there are fundamental limits in the material and energy interactions that are going on in the processes and so 3DP is unlikely to compete directly with systems like injection moulding and other techniques like casting. However, this does not necessarily matter as 3DP has an acknowledged place in the production of cost-effective customized products.[9] So in what ways might this potential scale up to impact upon development in the Global South?

The next print revolution

'Imagine, if you will, sitting down to your morning coffee, turning on your home computer to read the day's newspaper. Well it's not as

DOI: 10.1057/9781137365668.0006

far-fetched as it may seem.'[10] In 1981, the idea of reading the news online or downloading news articles everyday in order to print them out to read was as unlikely as 3DP in the home or community is now to many people; however, a magazine advertisement from 1981 displays a concept image familiar to us but alien then – the front page of a newspaper on a computer screen. In the 1980s, a browser webpage of images and text was not consumer-level because screens were typically black and limited to monochrome colours, and generally only single fonts were available with simple images or formatting. While text could be transferred over telephone lines, the process was slow and the cost of home modems was expensive. Yet, these practical limitations did not stop predictions of significant change in the future and, indeed, of a coming digital print revolution: 'Engineers now predict the day will come when we get all our newspapers and magazines by home computer'.[11] The 1981 feature reporter concludes that receiving the news online is unlikely to replace physical print because the transfer is too slow, the quality too low and the cost too high.

Fast forward to 2014. The news is indeed now mostly online. Although hard copies of newspapers continue to sell, this is a shrinking market and the major news corporations are disinvesting in physical media. Concerns about data transfer speeds and rates no longer apply with the mass adoption of home broadband and portable devices enabling formatting, images and even video to all be viewed with ease. There are many lessons in this example of the 2D digital printing revolution. The 2D digital revolution resonates with criticisms of 3DP today, which lacks speed, refinement, and economies of scale in comparison to factory-made products delivered via global supply chains.

So then a revolution proceeded full-tilt at the end of the nineteenth century with the mechanization of resource extraction, commodity production, and agriculture alongside the spread of information en masse. This was documented by many pioneering social scientists at the time: Auguste Comte, Emile Durkheim, Karl Marx, Max Weber, and Georg Simmel to name a handful. The second industrial revolution in the second half of the twentieth century was founded upon the invention of networked technologies, computerization, a rapid intensification of global mobility, and a worldwide knowledge economy.[12] A third print revolution, with 3DP in the vanguard, is now on the table and, according to commentators, has the following features that are relevant for development: efficiency dividends, environmental sustainability, reductions

DOI: 10.1057/9781137365668.0006

in production waste, off-grid networked 'smart' infrastructures, mass-customization, and reductions in transport emissions and congestion.[13] This time it is not only information being printed but objects as well.

The possibility of a looming third industrial revolution betokens a restructuring of the worldwide production and consumption system where most of the objects people currently use come from. As Anthropologist Anna Lowenhaupt Tsing pictures it, 'supply chain capitalism' is set apart from the weightless knowledge economy powering the Global North, but nevertheless relies on the Global South where industry and manufacturing happen.[14] Labour exploitation and environmental degradation are simply due course in the extraction and mobilization of resources and commodities, leading to varying nuances of austerity for people living in proximity to these activities.

Just as with the digitization of 2D print media, there are both evangelists and sceptics of 3DP. The former point to the swelling number of start-ups, patents, and ventures; the growth in use of 3D printers for rapid prototyping in design, engineering and architecture workshops and studios; the scope for customization, rediscovery of craft, maker communities and experimentation; and the growth in online repositories – both open and pay-walled – for uploading and downloading files. The latter point to the far from rapid printer speeds; the rough looking, comparatively expensive and flimsy objects printable on consumer home units; the inability of consumer printers to make objects with multiple materials, embedded electronics or metals; and the continuing economic merits of mass-manufacturing and supply chain capitalism. While time will tell who is right, there are indicators and forerunners now about what we call the 'Goldilocks Zone' of 3DP for development.

The 3D4D Goldilocks Zone

Pete Basiliere, research director at Gartner, says, 'We expect that a compelling consumer application – something that can only be created at home on a 3D printer – will hit the scene by 2016.'[15] While consultancies can provoke the markets, they predict to varying degrees as well as profit from certain outcomes, there is some balanced evidence and logic behind this forecasting. Just as in the search for life or interstellar exoplanets able to sustain human life, there is likely to be a Goldilocks Zone where the configuration of the elements of the 3DP ecosystem makes

DOI: 10.1057/9781137365668.0006

it 'just right' for development. A middle-ground printer that would be plausible for mass adoption and ultimately development as well is not yet incumbent. However, some features of this zone are possible to foresee now according to what we know about the development landscape and similar innovation revolutions such as mobile phones. What this brief review of the low and high ends of the 3DP ecosystem indicates is that there are many elements already on the market that point to a single suite of technologies, designs, materials, and infrastructures to support many of the claims being made about 3DP. Yet, some work is still required in imagining how 3DP might qualitatively impact upon development goals. While 3DP hype intimates that there is soon to be a Goldilocks Zone reached for the 3DP ecosystem there is also scepticism that the growth in users will continue to support research and development sufficiently enough to achieve economies of scale. The very low end – home printing – receives the most criticism on a handful of fronts spanning the areas this book identifies as key elements: technologies, designs, materials, and infrastructures. Sceptics highlight that the '3D printing revolution won't happen in your garage'.[16] Yet some garage-level innovations are making headway, such as work on an affordable domestic metal printer, and these are appropriate for the Global South.[17] Rather than criticize the 3DP ecosystem for its current limitations, we can try to imagine what would need to happen in order for the hype to be met across these dimensions. While there will most likely also be impacts that cannot be foreseen, it is undeniable that some baseline elements are anticipated for a Goldilocks Zone to happen relevant for 3D4D rather than consumer markets in the Global North.

First, *affordability* will be crucial for users with lower to the lowest incomes to be able to gain access to 3DP either individually or as community/shared facilities. The printers will need to be free from branding and cosmetic elements that add to the overall price but give no functional benefit. Designs will need to be bereft of patent and intellectual property costs so they can be shared and distributed widely. Materials will need to be sourced locally or recycled from waste, with industrial processing and transportation overheads limited.

Second, *flexibility* will be necessary for the conditions of the Global South. Printers will need to have longevity and replacement parts must be easily accessible to all. Printers would need to be flexible enough to be repaired on-site without service specialists. Repositories must offer a resilient catalogue of critical and non-critical designs for replacement

DOI: 10.1057/9781137365668.0006

parts and modules for larger more complex items. Materials would ideally be part of circular economies that tap into a number of different local and/or mixed sources.

Third, *simplicity* must be implemented across the ecosystem. Printers should not have regular upgrade cycles and the interfaces should not assume literacy/numeracy standards. Designs would not require Computer Aided Design (CAD) or coding skills or the operation of complex 3D scanners for reverse engineering. Materials would not require special handling facilities or training.

Fourth, to be firmly in the Goldilocks Zone, the 3DP ecosystem would require *scalability*. Scale requires a large user group for the printers. The crowdsourcing of designs drawing on many different experiences and material economies must provide enough information to satisfy demand.

Fifth, standardized *quality* must be mandatory for printers, materials, and designs. This facet extends to the durability of the prints, the longevity of the materials, and the care taken with the digital files available for the machine.

Once these five requirements are met, it is not difficult to foresee the viability of 3DP for development goals. Experts in 3DP are currently asking questions about when the Goldilocks Zone might be reached. In January 2004, an interdisciplinary group of leading researchers from science and engineering, management, economic development, and public policy came together to examine long term opportunities for manufacturing in the United States. One key issue they flagged in their report is the scope for 3DP to 'change the roles of traditional consumers and producers' wherein 3DP gives credence to economical low production volume manufacturing in the short term and in the long run even full end-user participation in the design of some more intimate products.[18] David W. Rosen, who participated in the workshop, estimates that the Goldilocks Zone printer (affordable, flexible, simple, multi-material) will emerge in less than ten years making 'the large centralized manufacturing facilities that predominate now... no longer be necessary'.[19] The development of the computer sector also offers some clues about what the Goldilocks Zone would be like.

One final thing that needs to be noted is that the Goldilocks Zone might not necessarily be reached through the mass adoption of home 3DP and this is where 3DP might diverge from the Internet or the mobile phone. One of the most influential people working in 3DP is engineer Richard Hague at the University of Nottingham's Engineering

DOI: 10.1057/9781137365668.0006

and Physical Sciences Research Centre (EPSRC) Centre for Additive Manufacturing. One of the misconceptions about 3DP that Hague thinks does not reflect reality is that every home will have a printer in the same fashion as digital paper printing. Rather, Hague suggests that 3DP is more likely to scale up in future in a similar manner to digital photography where print shops remain as the main medium for digital data to be converted into photographs. People might have access to portable devices, for instance 3D scanners, which support 3DP, just as portable digital cameras underpin photograph printing. However, through utilizing local suppliers with medium to high end printers, beyond extrusion technologies, the many unique benefits of 3DP that cannot be replicated elsewhere will be made available to the general public. These include geometrically complex designs, new materials and other innovations not even conceived of yet. So does the Goldilocks Zone point to a 3D printer that is ideal for development purposes?

Benchmarking 3DP

3DP is finding itself in the same headspace as technologies associated with the Industrial Revolution because it involves: technological aspects (factories, machines, materials) and social aspects (customization, division of labour, skills, crafts). What the gradual coalescence of the various elements in the 3DP ecosystem towards a Goldilocks Zone indicates is a singular opportunity for human development goals to be met by a grassroots innovation based on meaningful technology transferal between the Global North and the South, which trumps top-down development regimes through the promotion of open source, low-cost, and user-led circular economies. Before visiting these macro-societal impacts in more detail in the next two chapters on what 3DP changes, a number of studies are worth mentioning which point to the most likely configuration of a 3DP ecosystem in tune with the development goals of the Global South.

Benchmarking efforts are currently underway in order to provide a preview of the most likely candidate for a 3DP ecosystem that functions firmly within the Goldilocks Zone: one that is 'just right'. In a landmark effort, engineers D. A. Roberson and colleagues at the University of Texas printed the same file on five consumer and professional grade 3D printers, with costs ranging from US$1400 to US$20,900 according to third party standards of build time, material usage, dimensional

DOI: 10.1057/9781137365668.0006

accuracy and surface roughness as well as post processing, portability, and safety.[20] Interestingly, the consumer targeted Makerbot Replicator, based on the open source RepRap platform (more on this in a moment), ranked highly in their iterative ranking system for comparing desktop units, pointing to the advent of 'massified' 3DP. Of most interest is the scope for users to modify and customize the internal workings of the Replicator and its progeny in the same fashion as personal computers in the 1990s, a key reason for their possible proliferation in the Global South.

Industry consideration of [what] the ideal set of elements are for a 3DP ecosystem within the Goldilocks Zone is, as to be expected, a top priority for the leading incumbents and entrepreneurial start-ups. Intriguingly, the recent hype around 3DP has not been driven by the market, but rather the advent of an open source 3D printer from engineer Dr Adrian Bowyer and his colleagues from the University of Bath: the RepRap. Surprisingly candid about the reasons for making the RepRap open source, Bowyer is also a supporter of the 3D4D Challenge and the Reprap is the most likely candidate for achieving this aim.

Taking the last benchmarking exercise of currently available consumer printers further, tests of the economic viability of open source printers for the individual, household or community by B. T. Wittbrodt and colleagues at Michigan Technological University are positive about the potential massification of open source 3DP. They sifted out 20 files printable in thermoplastic from the 100,000 chosen in their benchmarking exercise from the open source Thingiverse repository. While a handful were novelty items, others had a clear utilitarian purpose applicable for everyday use in the Global South safety razors, food preparation tools, watchbands, callipers, shower heads, and protective phone cases. Of even more significance for 3D4D, this in-depth study clearly shows the cost benefits of distributed manufacturing on RepRaps, even including energy and feedstock costs: 'On average the products cost less than one dollar a piece to print. In comparison, online retail costs ranged from of US$300 to US$1900, averaging between US$15 and about US$100 per product'.[21] The authors also reject common concerns about the print quality and learning curves with design software; they note that all the designs were available ready-to-print from Thingiverse for no cost and that most quality concerns were cosmetic for the items they printed. Most valuably the authors conclude that a number of possible implications are: an expected rapid growth of 3DP through positive feedback

DOI: 10.1057/9781137365668.0006

loops; large-scale adoption and shifts to life-cycle thinking in consumption due to control over hyper-local manufacturing; growth of localized cottage industries in historically inaccessible high price items; and a revitalization of hands-on engineering-based education with no fear of patent, brand, or intellectual property violations. All of these trends are appropriate for 3D4D.[22]

'The productivity of hobbyists remains a potent force in technological evolution.'[23] What forecasting and benchmarking exercises demonstrate is that it is open source innovations led by informal developers and advocates that are most likely to provoke a systemic shift in the production and consumption of objects. In the case of the RepRap, proliferation in the developing world is likely to be considerably enhanced by the possibility of open source 'self-replication' capabilities of basic components powered by social networks and communities of grassroots innovators. In the next chapter, we turn to what 3DP will change about society and the various systems people rely upon for the objects they use in everyday life.

Notes

1 Woudhuysen (2013) 'Drilling into 3D printing: Gimmick, revolution or spooks' nightmare?'.
2 D'Aveni (2013) '3-D Printing Will Change the World'.
3 Birtchnell, Viry and Urry (2013) 'Elite Formation in the Third Industrial Revolution'.
4 Wallop (2013) 'With 3D printers, you can change the world (or at least mend a train set)'.
5 Giddens (2012) 'Re-industrialization in a New Electric Age'.
6 Giddens (2013) 'Lord Giddens on 3D Printing'.
7 Reeves (2013b) 'Jam Tomorrow: The Realities of 3D Printing Today'.
8 Science Museum (2013) 'Chris Tuck on Why 3D Printing Is Not 'Instant'.
9 Tuck and Hague (2006) 'The Pivotal Role of Rapid Manufacturing in the Production of Cost-Effective Customised Products'.
10 Newman (1981) 'Newspapers on Computers'.
11 Newman (1981) 'Newspapers on Computers'.
12 Urry (2007) *Mobilities.*
13 Rifkin (2011) *The Third Industrial Revolution: How Lateral Power is Transforming Energy, the Economy, And the World.*
14 Tsing (2012) 'Empire's Salvage Heart: Why Diversity Matters in the Global Political Economy'.

15 Rivera and van der Meulen (2013) 'Gartner Says Worldwide Shipments of 3D Printers to Grow 49 Percent in 2013'.
16 Whitwam (2012) 'Why the 3D Printing Revolution Won't Happen in Your Garage'.
17 Anzalone, Zhang, Wijnen, Sanders and Pearce (2013) 'A Low-Cost Open-Source Metal 3-D Printer': 807.
18 Shapira, Youtie and Urmanbetova (2004) 'Advanced Technology and the Future of U.S. Manufacturing'.
19 Shapira, Youtie and Urmanbetova (2004) 'Advanced Technology and the Future of U.S. Manufacturing': 42.
20 Roberson, Espalin and Wicker (2013) '3D Printer Selection: A Decision-Making Evaluation and Ranking Model'.
21 Wittbrodt, Glover, Laureto, Anzalone, Oppliger, Irwin and Pearce (2013) 'Life-Cycle Economic Analysis of Distributed Manufacturing with Open-Source 3-D Printers': 716.
22 Wittbrodt, Glover, Laureto, Anzalone, Oppliger, Irwin and Pearce (2013) 'Life-Cycle Economic Analysis of Distributed Manufacturing with Open-Source 3-D Printers': 719.
23 Doyle, Froede, Saint John and Devon (2010) 'In the Beginning Was the Noösphere: Community and Collaboration in Open Source Evolution of Technology': 7.

DOI: 10.1057/9781137365668.0006

3

What Does 3D Printing Change?

Abstract: *The structure of the worldwide production and consumption system that represents the status quo for making and moving most of the world's objects is undergoing extensive change due to innovations, notably 3D printing, which irrevocably alter the roles of producers and consumers. People suffering material poverty in the Global South are currently a crucial but underrated part of this system, through their endurance of inequalities and austerities in labour, environmental conditions and standards of living that would be intolerable in the Global North. In this chapter, Birtchnell and Hoyle examine changing forces in production and consumption arising from post-Fordist manufacturing methods that privilege mass-customization, made-to-order objects, and craft aesthetics. Moreover, producing-consumers – 'prosumers' – now demand accountability and transparency from their chosen methods of production, and they invest their identities in objects through their interactions as 'produsers'.*

Birtchnell, Thomas and William Hoyle. *3D Printing for Development in the Global South: The 3D4D Challenge.* Basingstoke: Palgrave Macmillan, 2014.
DOI: 10.1057/9781137365668.0007.

How the other half makes

The question of what 3DP will change is pertinent because, as Susan George writes in the Foreword to the reprint of her seminal work on world hunger, *How the Other Half Dies*, the poor:

> aren't really needed as consumers either. Classical economics tells us that capital must expand to ensure continued profits and that it does this by selling more goods to more people in increasingly remote corners of the world. An unblinkered look at present reality reveals, on the contrary, that 'the system' sees millions of people out there who are never going to become consumers. Whatever the efforts of the World Bank and assorted agencies, it's just too damned expensive to provide the infrastructures and create the jobs needed to draw them into the getting-and-spending mainstream.[1]

George candidly argues that despite ongoing rhetoric about lifting the Global South out of poverty through the mass conversion of its poor into consumers, there is little scope to do so within the current status quo that depends on inequalities to endure for continuing profit in factories, supply chains, and production networks. The challenges of creating a getting-and-spending mainstream in the Global South are too insurmountable at the moment to broach. In this chapter, we instead explore what 3DP will change about the current worldwide production and consumption system and elaborate how shifts of ideology around post-Fordist manufacturing and prosumerism intimate a revolution of a very different kind to other industrial epochs.

The division of labour

Anyone who has put their hands to making an object will realize that creating one from scratch requires many nuances of skill and expertise; the industrious balancing of physical energy and knowledge; various specific and general tools, accessories and facilities; and resources that can be gathered from the world around or fabricated synthetically. Objects that are not made in factories by machinery are brought to life from natural and synthetic elements by the careful handiwork of crafts-people and artisans, whether on a potter's wheel, a workbench, a hearth or an anvil.

What was so 'revolutionary' about the Industrial Revolution? This historic event represents the delinking of craft knowledge and skill from

DOI: 10.1057/9781137365668.0007

the production of objects, a process that gathered pace in eighteenth and nineteenth century Europe. All manner of efficiencies arose from innovative technologies as people took advantage of a great transformation: a surplus of resources due to international trade and the expansion of colonies.[2] Industrial 'engines' automated laborious tasks and allowed more objects to be produced quicker to the same level of workmanship regardless of the quantity. However, rather than freeing people from the task of making objects, machines divided the repetitive work amongst many others and created more work for operators of machines who had little or no craft or artisanal skill: the division of labour.[3]

So then, during the Industrial Revolution craft occupations were divided up into different functions managed by unique machines rather than by a single expert. The big mills and factories came to include in their ranks all manner of poor and unskilled people who would be recruited to work on production assembly lines that would move objects through the production process from stage to stage, often for long stretches and in unhealthy conditions. So the lives of those suffering poverty became entwined with mass-production.

Throughout the Industrial Revolution, the homeless, disadvantaged, indigent, and disabled were put to work as labourers within production networks of objects through an ideology of self-help. The reformer Samuel Smiles's book of the same name represents a key example, giving detailed biographies of the various innovators of the Industrial Revolution: Richard Arkwright (spinning frame), George Stephenson (railway) and James Watt (improved steam engine).[4] Social reformers saw this new energy of invention and machinery as a call to arms against malaise and listlessness and a campaign against poverty ensued, as the unemployed were entreated to follow the lead of captains of industry.

It was not only the production process that came to include the destitute in the Industrial Revolution, but also distribution. In the nineteenth century, production networks were largely simple affairs involving the carrying of stock from manufacturer to market. In larger markets, cities for instance, large cohorts of itinerants played a crucial role in these local distribution networks, which in turn provided them with piecemeal livelihoods. The identities of these ad hoc workers became tied to the objects they would distribute in this new production system as the reformer and journalist Henry Mayhew colourfully describes in his book *London Labour and the London Poor.*[5] Characters in this book are known by the objects they carry about to sell: Street-Sellers of Crockery-Ware, Long-Songs,

DOI: 10.1057/9781137365668.0007

Dogs' Collars, Walking Sticks, the Lucifer Match Girl, the Street Comb Seller, the Blind Boot-Lace Seller, the Street-Seller of Nutmeg-Graters.

The division of labour – first across the counties of England then the countries of Europe and finally across the world – has continued to be the dominating mode for object production par excellence. What does 3DP change about this arrangement? This new process changes again this link between the production process and the person, making it possible for people who have never before undertaken the creation of an object to produce one personally without the collective enterprise and investment in expensive collective machinery. Crucially, 3DP also allows the personal distribution of objects. This represents a reunification of labour into a single fluid process, yet without the component of artisanal skill that has limited object creation to a few 'makers' since prehistoric times. The enduring role of poverty in the division of labour suggests that the digital reunification of labour in 3DP is sure to impact upon inequality. This chapter considers what 3DP will change about the making of objects in light of the intimate relationship between poverty and production.

Global production networks

In the second half of the twentieth century, information and communication networks and offshoring have allowed the division of labour to span whole continents in order to recruit workforces in poverty to make objects in radically divergent circumstances from the consumers of these same objects. Many of the benefits of this worldwide system of production and consumption are out of sight and out of mind in the Global North because they happen in other 'poorer' places; however, these same benefits of place for consumers in the Global North exclude many in the Global South from involvement in this system due to the need for them to remain poor so that profits stay high.

'No matter how mobile some firms are, no matter how free to migrate some workers might be, no matter how much communication advances have shrunk the globe, place still matters for production, reproduction and consumption.'[6] From the 1980s onwards, social scientists turned their attention to the end of what they understood as 'organized' capitalism: the competition within distinct countries of organizations for profits derived from the provision of consumers with both production, marketing, and management jobs and consumer

DOI: 10.1057/9781137365668.0007

objects. Capitalism became disorganized, instead deploying complex networking strategies in order to move operations away from distinct regions altogether.[7] What scholars concerned with this phenomenon are cognizant of are the continuing territorial and geographical natures of these disorganized networks, reminding us that production 'had never been abandoned but merely outsourced'.[8] Global production networks (GPNs) involve regime-driven factories in the Global South with youthful labour cohorts, container ships with a handful of crew, machine-run cargo ports, cavernous warehouse inventories and so on. Attention to this new system built around GPNs through many supply chains summons up renewed thought about the Fordist production and consumption system that dominated in the late twentieth century and formed the 'back end' of the worldwide finance and market trading system of the early twenty-first century. However, the face of this system designed to shift mass-made objects rapidly around the world to the Global North and to select clusters in the Global South is an entirely different beast.

All manner of 'things' flow through GPNs ultimately to the homes of the Global North and inevitably to landfill or back to the Global South as waste exports. These are part of social settings and practices that encourage cyclic renewals of consumption without consideration of the production activities that support them. Of most interest to those considering the inequalities of this system are the deep risks and exploitation, both inherent within the workings of disorganized supply chain capitalism. 'In the current days of global supply chains and multinational companies, mistakes can occur at any point in the value chain, as products are designed, manufactured, and sold around the world.'[9] To be sure, many of the incongruities with offshoring occur because of the tenuous fabric of this divided system, composed as it is of such distant actors bound by undisciplined regulations. Moreover, much of the cost-effectiveness of this system is driven by all sorts of cuts made elsewhere from consumers, not only in product longevity and quality, in being 'made to break', but also in the conditions of the region where production happens: pollution control is lax, labour wages are low, occupational health and safety and emissions management are neglected.[10] It is perhaps this last point that is most significant in underlining the inequalities of this system.

It is not only the conditions of workers or the quality of objects that make offshoring questionable, but also the 'embodied energies' in these objects: emissions of greenhouse gases that move across territories invisibly. Geographer Luke Bergmann talks of an alternative method of visualizing these inequalities through depicting chains of carbon embedded in the goods moved upstream from the Global South where they are produced, to the Global North where they are consumed.[11] In effect, this alternative method associates the consuming region with carbon emissions that occurred far beyond its borders regardless of the emissions spent at the time of production – the current method for assessing global emissions responsibilities that site them in manufacturing regions.

So then a major issue for supply chain capitalism is that the system's track record in boosting the living standards of the poor majority in the Global South is dire. The minority who disproportionately benefit from the system in the Global North do not find it to be expedient to tackle poverty in places that appear to have little physical relationship to them despite being the ultimate source of the objects they use every day. Only the labels they bear remind them of their international provenance. The back-end of global commodity chains thus depends on the international dimensions of austerity to remain unequal in order to remain profitable. This state of affairs is brought home by the shift of the bulk of the world's poverty to Middle Income Countries (MICs) despite some of these countries even becoming aid donors themselves as part of a new geography of global poverty.[12] Senior Economist Francesca Beausang details two reasons the MICs have not shed their poverty burdens with their inclusion into world production and consumption.[13] First, domestic inequality hobbles any meaningful efforts towards development. Second, ideas and innovation are still held and produced in the Global North.

Alongside the various strategies that continue to make supply chain capitalism profitable – the most questionable being 'planned obsolescence' – there are also the risks inherent in a system made up of many tentative links in long chains spanning across territories and environments. When catastrophes occur this is when 'the smooth operation of the space of flows is disrupted, and when the often-invisible networks of mobility are made visible'.[14] So then the GPNs, which power the world economy are far different to the consumer cultures they feed products into. The geopolitics of the GPNs described so far in this chapter are hazy to most consumers within the system. There are indications, however, that change is afoot in how consumers relate to producers.

DOI: 10.1057/9781137365668.0007

Post-Fordism

In his book *The Third Industrial Revolution*, Social Scientist Jeremy Rifkin considers what he calls the 'morph' from the industrial to the collaborative era over three successive stages, a transition equating to the end of mass-production, elsewhere termed 'post-Fordism'.[15] The first Industrial Revolution is well known to all aware of early social science writing on the division of labour – the rapid shift from domestic craft and cottage industries to large urban factories in the Global North that saw the development of the 'American model' of mass-production: separation of fabrication from assembly, big batches, inspection and standards, low-skill, low-wage operators, simple assembly.[16] The second industrial revolution of the twentieth century was characterized by Henry Ford's legacy of nation-building manufacturing and social reform that saw the working poor in the US become middle-class and the labour workforce widen across nations to take on board those suffering poverty elsewhere as labourers. As Historian David Nye makes clear in *America's Assembly Line*, mass production also drove mass consumption as factory workers were motivated to purchase the vehicles they worked on and were paid more as a consequence (the symbolic US$5 a day) of their promise to spend.[17]

The propulsion towards full factory automation and the ultimate replacement of human labour from manufacturing lapsed in the mid-twentieth century in favour of GPNs that relied instead on inequality across regions to fuel factory labour from vast cohorts of poverty sufferers in the Global South. Rifkin maintains that the transition to a more equal era where machine replaces human labour has until now been undone by the conundrum that increasing productivity in the form of intelligent technologies, robotics, and automation dampens demand for consumer objects by lowering incomes and, more seriously, pushing labour recruitment offshore.

The move of manufacturing offshore arose with the discovery of cheap energy, particularly oil, stimulating a transition to far-flung manual labour rather than a transition to total automation.[18] So then in Rifkin's second revolution an energy bonanza spring-boarded a rapid upscaling of transportation and infrastructure systems. This global boon derived from abundant resources shortened distances and heightened inequalities: 'manufacturing today is carried out to a substantial degree in emerging and developing economies, creating a changing and complex landscape of global production'.[19] So a final stage in the second industrial

DOI: 10.1057/9781137365668.0007

revolution is the offshoring of labour, taxes, carbon emissions and environmental and moral responsibilities through complex network technologies.[20]

One key change 3DP brings is the emergence of the digital into the real world as tangible objects appear on the printer's build tray, unlike previous technology innovations (the Internet or the mobile phone), which were virtual and intangible. Some assessments of 3DP's emergence predict it playing a central role in the rise of peer-production movements, which 'value democratic aspects of voluntary horizontal cooperation, and prefer creativity and happiness to money and careerism' and display 'concern for ecology, antipathy towards consumerism, and care for the poor and the Third World'.[21] Others firmly locate 3DP within the current political and economic system and even foresee a manufacturing renaissance in the nearer 'back end' stemming from the return of production to the Global North through advanced technologies. However, the 'front end' could also benefit from 3DP according to the CIO of supermarket chain Tesco, Mike McNamara, who envisions retail decor harbouring printing technology in-stores to fully realize bringing just-in-time business models down to minutes rather than days: 'To see this as a fight between physical and digital is to see it all wrong. Already today 65 percent of our online orders are click and collect'.[22]

So as an indicator of a post-Fordist shift, 3DP does represent a significant divergence from the existing system – hence its attribution to a 'new' or 'third' industrial revolution – for a number of key reasons, which management expert Barry Berman at Indiana University summarizes. Of these, it is cost effectiveness when creating custom products in small quantities that makes 3DP stand above other similar innovations. The advantages of 3DP in comparison to other manufacturing technologies are:

> No need for costly tools, molds, or punches; no scrap, milling, or sanding requirements; automated manufacturing; use of readily available supplies; ability to recycle waste material; minimal inventory risk as there is no unsold finished goods inventory; improved working capital management as goods are paid for before being manufactured.[23]

Custom products that are made to order are appropriate for developing countries' markets where current factory-produced objects fail to engage – the much-vaunted informal markets at the Bottom of the Pyramid (BOP), more on this is discussed later in Chapters 5–6 of this book.

DOI: 10.1057/9781137365668.0007

Governments certainly appear to foresee 3DP representing a boon to ailing manufacturing regions in the Global North. 3DP features in a variety of assessments scoping pathways to bring the back-end (manufacturing) back on a par with the front-end (services) in these regions. UK projections observe growth into the 2020s with stimulation from 'manufacturing on demand' disrupting the twentieth century 'mass market' model.[24] According to a significant report on the future of manufacturing by the UK Government Office for Science, 3DP affords 'designs to be optimised to reduce waste; products to be made as light as possible; inventories of spare parts to be reduced; greater flexibility in the location of manufacturing; products to be personalised to consumers; consumers to make some of their own products; and products to be made with new graded composition and bespoke properties'.[25] Likewise for the rest of Europe, 3DP offers 'customized production in much smaller quantities' with consequent savings in energy and raw materials and opportunities for SMEs.[26] Rather than representing a U-turn from services in the Global North, this would be an extension of the shift to a primarily 'front end' society outlined earlier: in this future front-end service sectors would also involve back-end production in one package.

'The practices of distributed manufacturing... mark a progression on the post-Fordist trajectory... beyond interconnected organisational structures and into a rhizomic network of globally dispersed individuals.'[27] Distributed manufacturing is not simply the demise of the mass-manufacturing factory. 3DP advocate Chris Anderson points out that networks of dispersed individuals already access industrial-class warehouse-sized data factories remotely when completing a basic Google search and, therefore, it is not such a stretch to imagine individuals accessing digital fabrication factories in the same manner.[28]

Since the mid-1990s, some general dimensions have been agreed on that define post-Fordism centring on flexibility.[29] First, new microelectronic technologies, the latest being 3DP, give the consumer the power to disrupt the assembly line logic of standardized and uniform products.

Second, systems emphasizing economies of scope with made to order production ideologies. Indeed, services such as Dutch-founded Shapeways already offer consumers a sophisticated range of objects from their 3DP 'factory' in Queens, New York, which houses 50 industrial printers.

Third, there is the scope for leaner, flatter and even leaderless organizations accompanied by unique financial instruments not excluding virtual

DOI: 10.1057/9781137365668.0007

or real bartering-style marketplace models. 3DP companies seem to be an extension of Information and Communications Technology (ICT) companies with their lack of stratification and campus-style organizations involving only a few founding members and many sub-contractors and independent associates.

Fourth, uniquely divergent social and work relationships, autonomous from national systems are far more sensitive to networks.[30] 3DP maker movements are a most appropriate example of the digital spilling over into the physical world and shaping alternative social values and ways of living.

It is clear that most 'predictions are based on a technological analysis that compares the abilities of the new 3D desktop printers with the industrial capabilities of mass production'.[31] The idea of post-Fordism predates popular interest and awareness of 3DP and this arises as a surprising real-world example of many of the ideas within this turn of thinking. The dissolution of the factories might not be on the horizon; however, there is consensus that 3DP heralds a revolution of sorts for the Global North. But where is the Global South in these discussions? And what role does 3DP play in rectifying the imbalances between regions that have accelerated dramatically since the Second World War and the great transformation of containerization, supply chain capitalism, consumerism and so on?[32] To answer these questions about what 3DP changes about development, we must first understand a seam of thought within post-Fordism that is the trigger for both the hype around 3DP and expectations of an imminent industrial revolution: prosumerism.

Prosumerism

So far the discussion has catalogued the real and speculative interventions of the 3DP ecosystem in the – purposefully polarized in many accounts – back-end of the current system of production (supply) and the front-end of consumption (demand). Printer technologies still need to be mass-manufactured and sold on global markets; however, 3DP allows circular economies to look more like cosmopolitan consumer cultures by marrying informal economies up with modernized products and objects through informalities: barter, sharing, reverse engineering, and copying. After describing these two theatres, we concluded that 3DP is now heralded as a reversal of a long trend in the Global North's economies away from the domestic, localized production of objects towards strict consumption from a handful of manufacturing regions in order to satisfy a core social drive: the lust for the cyclic accumulation and disposal of objects. The discussion kept pace with the manifestation and consequent influence of post-Fordism as a means to encourage a manufacturing renaissance in the Global North to bring production back on a par with consumption and away from the geopolitical and energy risks of supply chain capitalism. In this section, we visit a second change relevant to 3D4D, similar but in the opposite direction to post-Fordism and also spanning both theatres of supply and demand: the notion of the prosumer, the consumer who is also a producer.

'Mass customization' is the latest trend in manufacturing and a response to demand for product variety and the scope for personalization inherent in the democratization of digital data. But mass-manufacturing using injection moulding and casting techniques is not very suitable to this paradigm shift.[33] This shift in demand signals a new niche that 3DP is able to partly fill. Prosumers empowered with 3DP bulldoze the currently complex system of making things: patenting an idea, romancing venture capitalists, conducting market research, manufacturing costly prototypes in producer regions, bulk manufacturing, freight and inventories, sales and retail contracts and repeated oceanic shipping throughout the whole process. Stephen Fox with the VTT Technical Research Centre of Finland summarizes this potentially open prosperity in a disarmingly simple premise: 'anybody anywhere can make use of digitally-driven manufacturing technologies to produce original products' and because they are extremely portable they can spread even to landlocked African countries including 'Burundi, where up to 75% of the cost of goods can arise from transportation'.[34]

A key change the 3DP ecosystem brings to the table is to the domestic consumer and the degree to which they play an active role in their own consumption of produced objects as prosumers. 3DP is just one of a number of innovations that indicate a prosumer revolution: crowdsourcing, peer banking, creative commons, and 'mashable' digital data.[35] Production and consumption are part of a formal economy, where people get paid salaries or welfare, pay taxes, and spend their earnings on consumer commodities; this system is skewed across different regions through supply chain capitalism, which currently frees the domestic consumer from the labour of object production: resource extraction, transportation, storage and distribution, until the point of sale either

DOI: 10.1057/9781137365668.0007

online through the Internet or through a specialist retailer or aggregator of many objects – a supermarket or mall.[36] The 3DP hype described earlier focuses most fervently on the potential for change in how the domestic consumer relates to formal systems of consumption and its separation from the production process – the divorce from, rather than division of, manual, craft labour that went on before the Industrial Revolution.

It is not an exaggeration to state that consumers in the Global North no longer (or at least to a quantitatively marginal degree) participate in the processes of production within the global trade system. Political theorists George Ritzer and Nathan Jurgenson deftly put into perspective the dramatic shift that has taken place in the way production and consumption function together in the same system. Surprisingly, they suggest that despite appearances consumers began as labouring prosumers in early factory-focused capitalism (Ford's consumer workforce) and, in recent times, have returned to become labouring participants in the production process through 'self-service' fast-food business models and more uncontrollably (and unpredictably for business) in Web 2.0. This is a new way of understanding, and depolarizing, production and consumption for in 'producer and consumer capitalism, corporations are likely to exert great control over the production and/or consumption of content (goods and services), but in prosumer capitalism companies are more likely to stand back and to meddle less with the prosumers who are producing and consuming the content'.[37]

The most radical version (for corporations at least) is 'commons-based peer production' where manufacturing is relocalized by prosumers through 3DP, reducing the need for 'an assembly line, not to mention the reduction in carbon footprint due to less transport' and producing 'economies of scope' rather than scale.[38] By moving the focus of consumption to scope rather than scale, users are no longer simply an audience for the performance of globally circulating objects, a performance staged by marketing and advertising; instead, prosumers are put on the stage as fellow performers engaging in 'playbour' through their interaction with all manner of digital data in everyday life.[39] This latest manifestation of prosumption through data accumulation and participation naturally extends to 3DP, which makes data physical. Hence, the hype around home printers appears to be the ultimate manifestation of the prosumer according to some commentators.

Focusing on the scholarly considerations of 3DP's potential for social change through its affect on domestic consumers is Sociologist Mike

DOI: 10.1057/9781137365668.0007

Michael's linkage of predictions of the eventuality of the 3DP ecosystem as a dominant system in society and of hopes for the democratization of plastic and indeed 'plasticity': the manipulation and transformation of this very special and malleable human-made material into all sorts of useful and not so useful items.[40] While tending to ignore the diversity of 3D printable materials available at the high-end (in additive manufacturing) and unlikely to become available to the domestic consumer printer, Michael makes a number of crucial points in theorizing what 3DP could change about the formal nature of plastic materialities and possible social transformations to the worldwide production and consumption system.

The first point for Michael is that 3DP hype weaves into prosumer narratives and imaginaries about what he terms 'everything-ness'. Speculation on a 'factory-in-a-box' and such ilk feed into alchemistic desires for matter to heed the many aesthetic and practical demands on its form by consumers, which supply chain capitalism, despite its delivery of a menagerie of objects, remains unable to state due to its contrived and real limitations. Various publicly shared fictions are instrumental in supporting 3DP hype about everything-ness.[41] Of all the materials available to the domestic consumer, it is disposable and cheap plastic objects that feature the most in these 3DP imaginings.

The second point Michael contributes singles out the 'easy-ness' for prosumers of the proposed 3DP ecosystem and its transcendence of the labour in the craft or manufacture of objects: skills machine labour has made invisible to consumers in the Global North through the movement of mass-manufacturing 'offshore' to the Global South and elsewhere.[42] Domestic users anticipate greater levels of ease that go beyond the physical limits of supply chain capitalism and mirror the ease of the digital world. Undoubtedly the ease with which virtual coded 'matter' can be transformed readily and [albeit by experts] rapidly, in some cases marginally or totally replacing physical objects (clocks, newspapers, or writing pads), holds a stake in desires for easy-ness in the home and other informal domains.

Yet, for others it is this very 'everything-ness' coupled with 'easy-ness' that threatens society. Economist Thomas Easton writes of the perils of domestic 3DP for the 'economy of trust' that governs the value and availability of objects to consumers. For Easton the idea of prosumers is profoundly negative, as it harks back to the early days of the Internet and the bloom of file sharing and peer-to-peer piracy across digital networks that robbed some producers, such as musicians, of their rights to their

DOI: 10.1057/9781137365668.0007

products, which they enjoyed before online networks. For Easton, the ease of 3D scanning and printing any object imaginable triggers alarm bells for the sanctity of material values beyond their core constituents: the labour spent in craft, their age, or the rarity of the object's resources. According to Easton, the possibilities for illegally sharing digital designs of antiquities, fossils, technical tools, jewellery or weapons poses a severe threat to formal economies and the stability of the global financial system itself, which remains tied to consumer capitalism. 3DP is a disruptive technology and these 'change the rules ... destroy businesses ... make industries and jobs obsolete ... even cut the funds available to governments to support schools and maintain roads or ... force government to find the funds to pay for new services'.[43] Here the idea of the 'prosumer' is not necessarily an empowering development as consumer identities are tied to the process of becoming a commodity to market: an intangible object for profit.[44]

So a prosumer revolution could on the one hand rebalance the global economy, now regionally skewed between consumption and production, or on the other hand throw the Global North into disarray through the dissolution of trust. Yet, what might a prosumer shift hold for the Global South, where societies are already lacking in 'trust', where manufacturing and physical labour continue to grow, where infrastructures are lacking and taxpaying unenforced, and where employment and commodities are for the most part managed through an informal economy?

First world problems

Many commentators insist that at root '3D printing will only begin to replace mass production if it becomes cheaper, of higher quality, or can offer customers and businesses something significantly different to traditional manufacturing processes'.[45] Heretofore, there was no alternative to the dominant system of worldwide production and consumption, which this chapter abstracts as the 'back end' and 'front end' of globalization. Sceptics of 3DP's potential to invoke social change are adamant that the innovation is unlikely to be able to replace globally distributed Fordist mass-manufacturing to any meaningful degree beyond obvious industry prototyping needs. The tongue-in-cheek Internet meme 'first world problems' (and accompanying hash-tag #firstworldproblems) includes examples of common sufferings: no Wi-Fi in a cafe, slow Internet access,

DOI: 10.1057/9781137365668.0007

no parking close to the front-door, sunburn from the beach, too many traffic lights, dishwasher malfunctions, and so on. Many of the criticisms levelled against 3DP's potential for social change can be understood as also suffering from this issue when transposed to the Global South.

Evidence adduced to support the argument of 3DP's hype points to a number of currently unredeemable features, making the achievement of scale unlikely in the Global North. First, the cosmetic shortfalls and visual flaws of low-end plastic prints make market penetration of many objects discountable. Second, is the cost of materials without the benefits of bulk manufacturing. Third, are the unintuitive software and printer interfaces with their file format issues and non-conformity across brands. Fourth, there is the inability to compete with the allure of unboxing cyclic streams of cheap products buttressed by the sheer imaginative force of global marketing companies in concert with canny corporate product designers.

These are all indeed convincing hurdles; however, in defence of the 3DP ecosystem there are resolutions that make it appropriate for social change in the Global South. It is problematic to compare singular objects currently more akin to individual artworks or curios to runs of thousands, if not tens of thousands, of near-identical objects unleashed on global markets annually. As well, the early computer ecosystem met with similar responses: cosmetics, cost, complexity, and consumer caprice. What these points allude to is that it is not enough to adopt an unimaginative approach to foresight on 3DP's impact on social change.

Indeed, science fiction writers Cory Doctorow, Ian MacDonald, Charles Stross, and Neal Stephenson all evince compelling visions of possible future worlds where 3DP plays an integral part. In one recent exercise, these inputs were blended into two core hypotheses on 3DP in the year 2030 for foresight on degrees of individual engagement and privatization. This study broadcast across different scales (domestic, corporate, community, and technical users); a range of settings (home, shop, library, and firm); and newly adjusted social practices (off-grid recycling of 3D printed objects in the home, mass-customization giving retail consumers custom objects, utilities-style feedstock delivery, and an economic bubble around 3DP from too much hype).[46] Other blends of science fiction and expert visioning offer even more complicated models about how 3DP will transform society. For instance, imagining social production causing a business landscape dominated by small firms and entrepreneurs with no room for major conglomerates.[47]

DOI: 10.1057/9781137365668.0007

There is, however, a far more compelling point to consider that Geographer Neil Coe and colleagues highlight: 'regional development does not take place on a level playing field', for the most part because of the inherent inequalities in the balance of power latent within global production networks.[48] Conditions in the Global South are radically different for the majority of the people there than in the Global North, simply due to the overwhelming cohorts suffering poverty: The World Bank estimated in 2010 that there were 400 million in extreme poverty in India alone.[49] This means that many of the assessments of the opportunities and limitations of 3DP listed above are merely 'first world problems' unlikely to be relevant in the Global South and requiring an extended modelling of 'rapid prototyping for the masses'.[50]

Notes

1 George (1986) *How the Other Half Dies: The Real Reasons for World Hunger*: xvii.

2 Polanyi (2001) *The Great Transformation: The Political and Economic Origins of Our Time.*

3 Durkheim (1997) *The Division of Labor in Society.*

4 Smiles (1866) *Self Help.*

5 Mayhew (1851) *London Labour and the London Poor.*

6 Castree, Coe, Ward and Samers (2004) *Spaces of Work: Global Capitalism and Geographies of Labour.*

7 Lash and Urry (1994) *Economies of Signs and Space.*

8 Tsing (2012) 'Empire's Salvage Heart: Why Diversity Matters in the Global Political Economy': 39.

9 Bapuji (2012) *Not Just China: The Rise of Recalls in the Age of Global Business*: 154.

10 Slade (2006) *Made to Break: Technology and Obsolescence in America.*

11 Bergmann (2013) 'Bound by Chains of Carbon: Ecological–Economic Geographies of Globalization'.

12 Sumner and Mallett (2012) *The Future of Foreign Aid: Development Co-operation and the New Geography of Global Poverty*: 3.

13 Beausang (2012) *Globalization and the BRICs: Why the BRICs Will Not Rule the World for Long.*

14 Cresswell and Martin (2012) 'On Turbulence: Entanglements of Disorder and Order on a Devon Beach': 520.

15 (2011) *The Third Industrial Revolution: How Lateral Power Is Transforming Energy, the Economy, and the World.*

DOI: 10.1057/9781137365668.0007

16 Stahl (2013) *Worker Leadership: America's Secret Weapon in the Battle for Industrial Competitiveness*: 13.
17 Nye (2013) *America's Assembly Line*: 30.
18 Urry (2013) *Societies Beyond Oil: Oil Dregs and Social Futures.*
19 Coe and Hess (2013) 'Global Production Networks, Labour and Development': 1.
20 Urry (2014) *Offshoring.*
21 Rigi (2013) 'Peer Production and Marxian Communism: Contours of a New Emerging Mode of Production'.
22 Neagle (2013) '3D Printing: Good for Retail? Tesco CIO Says So'.
23 Berman (2012) '3-D Printing: The New Industrial Revolution': 158.
24 Foresight (2010) 'Technology and Innovation Futures: UK Growth Opportunities for the 2020s': 2.
25 Foresight (2013) 'The Future of Manufacturing: A New Era of Opportunity': 19.
26 European Commission (2012) 'Industrial Revolution Brings Industry Back to Europe': 5.
27 Richardson, Elliott and Haylock (2013) 'This Home is a Factory: Implications of the Maker Movement on Urban Environments'.
28 Anderson (2012) *Makers: The New Industrial Revolution*: 58.
29 Jessop and Sum (2006) *Beyond the Regulation Approach: Putting Capitalist Economies in Their Place.*
30 Jessop and Sum (2006) *Beyond the Regulation Approach: Putting Capitalist Economies in Their Place.*
31 Ilan (2011) 'From Economy of Commodities to Economy of Ideas: Hardware as Social Medium': 47.
32 Birtchnell (2013a) 'Fill the Ships and We Shall Fill the Shops: The Making of Geographies of Manufacturing'.
33 Hu (2013) 'Evolving Paradigms of Manufacturing: From Mass Production to Mass Customization and Personalization'.
34 Fox (2013) 'Open Prosperity: Breaking Down Financial and Educational Barriers to Creating Physical Goods': 23.
35 Ramos (2013) 'The Future of Power in the Network Era'.
36 Birtchnell and Urry (2014) 'The Mobilities and Post-Mobilities of Cargo'.
37 Ritzer and Jurgenson (2010) 'Production, Consumption, Prosumption: The Nature of Capitalism in the Age of the Digital "Prosumer"': 32.
38 Kostakis (2013) 'At the Turning Point of the Current Techno-Economic Paradigm: Commons-Based Peer Production, Desktop Manufacturing and the Role of Civil Society in the Perezian Framework': 181.
39 Beer and Burrows (2013) 'Popular Culture, Digital Archives and the New Social Life of Data'.
40 Michael (2013) 'Process and Plasticity: Printing, Prototyping and the Process of Plastic'.

DOI: 10.1057/9781137365668.0007

41 Birtchnell and Urry (2013) '3D, SF and the future'.
42 Urry (2014) *Offshoring*.
43 Easton (2011) 'A Recession in the Economy Trust': 160.
44 Comor (2010) 'Contextualizing and Critiquing the Fantastic Prosumer: Power, Alienation and Hegemony'.
45 Sissons and Thompson (2012) 'Three Dimensional Policy: Why Britain needs a policy framework for 3D': 17.
46 Birtchnell and Urry (2013) '3D, SF and the Future'.
47 Potstada and Zybura (2014) 'The Role of Context in Science Fiction Prototyping: The Digital Industrial Revolution'.
48 Coe, Hess, Yeung, Dicken and Henderson (2004) '"Globalizing" Regional Development: A Global Production Networks Perspective': 481.
49 Olinto and Uematsu (2010) 'The State of the Poor: Where are the Poor and Where are they Poorest?'.
50 Gordon (2011) 'Rapid Prototyping for the Masses'.

DOI: 10.1057/9781137365668.0007

4

What Does 3D Printing Change about Development?

Abstract: *Transfers are a major tool in relationships between aid giving and aid receiving regions. Transfers of technology offer solutions to many major development goals. In this chapter, Birtchnell and Hoyle focus on the changes that transfers of 3D printing will invoke in development discourse and policy with the ultimate goal of overall poverty alleviation. Niche grassroots level innovations are a crucial element in community action against material poverty and 3D printing intimates a different approach to furnishing the bottom of the pyramid with objects of use that could empower their lives.*

Birtchnell, Thomas and William Hoyle. *3D Printing for Development in the Global South: The 3D4D Challenge.* Basingstoke: Palgrave Macmillan, 2014.
DOI: 10.1057/9781137365668.0008.

Top-down development

> What would happen if we took 3DP and applied it to the developing world? A world where Internet-based retail is still in its infancy, but connectivity is growing. A world where the traditional supply chain, logistics, and delivery can take weeks, even months. How could 3DP impact upon the developing world if it could be used to support local manufacture, for agriculture, education, healthcare, fair trade, or simply just to provide access to some basic consumer goods and hardware.[1]

Just as the Internet changed the world in the 1990s, 3DP is set to change the world again. As described earlier in this book, the technology is revolutionizing the way that we make products by bringing the factory into the community and allowing computers and the Internet to become the new conduit for skills, innovation, and creativity. 3DP's applications across a range of industrial sectors, from aerospace, architecture, and automotive industries through to medical and dental implants, prosthetics and rehabilitation aids, demonstrate its flexibility and diversity. The technology is also finding applications in the consumer goods sector, as a new way of retailing products that are made to order, with little if no stock holding and no waste. How might the 'other half' lacking access to the conveniences, comforts, and cleanliness of the Global North perhaps be reconciled with some of these staples through 3DP?

In Chapter 2, we considered the 3DP ecosystem and established that a Goldilocks Zone for 3DP was likely to be entered in the next few years or at the most decades, delivering a vanguard for the mass adoption of this innovation. Chapter 2 examined how 3DP is now cascading down from professional and industrial users to the local community and home-based 3DP, which is set to revolutionize the way goods are purchased by allowing digital data to be downloaded from the Internet for home 'manufacturing', much in the same way as we download music and films today. 3DP has the potential to revolutionize the way we manufacture, consume, and live in our material worlds and this means new ways of conceiving of production and consumption with radically altered global production networks (GPNs). In Chapter 3, the impacts of this convergence between the low and high ends of the 3DP ecosystem were assessed in their likelihood to inculcate a measurable change to how objects within the world economy are made, moved, and managed within the 'back end' of supply chain capitalism. The 'front end' of shops, marketing, brands and so forth also stand to confront significant

DOI: 10.1057/9781137365668.0008

change should the 3DP Goldilocks Zone be realized. Bringing these two ends together, two key themes of post-Fordism and prosumerism were reviewed. This shift, considered in Chapter 3, is likely to be felt in both the back end (post-Fordism) and the front end (prosumerism) of this system. However, the novelty of 3DP is not done justice by comparisons to the existing system of worldwide production and consumption – the 3DP ecosystem fails on many accounts in regard to cost of single items, quality, speed of production, and economies of scale.

In the Global North, many of the more profound ramifications of 3DP are neither scalable nor realistic in comparison to mass manufacturing, given the many benefits of the current system and its incumbent power and support. Filling large factories with rows of 3D printers in order to make objects locally in the developed world is never likely to be competitive with supply chain capitalism. And mass-made objects with a century of industrial research and development behind them will also be more likely to satisfy consumers' psychological and materialistic desires, unless 3DP is able to attain standards greater than today's printed objects beyond moving parts: mixed materials, multiple colours, smooth and 'finished' textures, printed circuit boards, and so on. That there will be a substitution of mass manufacturing by 3DP stretches beyond the argument of this book. Rather we suggest that it is in the Global South where the 3DP ecosystem comes into its own, where large numbers of people get by outside of the current worldwide production and consumption system, and where the novelties and benefits have the potential to be cumulative once 3DP is introduced through tactical and sympathetic development action.

But development action between the Global North and South is far from a simple affair. Resolutions to address global inequalities are the remit of multilateral organizations, most popularly the International Monetary Fund (IMF), the Organization for Economic Cooperation and Development (OECD), the United Nations Industrial Development Organization (UNIDO), the World Bank Group, the World Health Organization (WHO), and the World Trade Organization (WTO). This top-down approach to development is the source of many reports and policy recommendations and, at the end of the twentieth century, produced the MDGs – a concrete commitment to multilateral action to reduce poverty.

The MDGs were a 15-year strategy to overcome extreme and degrading poverty and the eight goals aspire to address the overarching

DOI: 10.1057/9781137365668.0008

challenges the world's poorest face. The signatories to the MDGs recognize that the developing world's poverty traps are a problem for the stability of the world's financial markets and economic growth. Eight goals were laid down itemizing action on poverty and hunger, universal primary education, gender equality, child mortality, maternal health, disease, environmental sustainability, and global partnerships for development. The end of the tunnel is in sight for this 'big push' to end inequality by 2015; however, the degree to which the goals have been met is unclear.[2] Looking back on the tenure of the MDGs, there is a sense that more remains to be done on the ground in the Global South. The UN is already voicing its support for grassroots approaches to development, as 'creativity and innovation are a natural resource in which every country and every community is potentially rich'.[3] We suggest it is in this space that 3DP has the scope to provide a pragmatic solution that is 'bottom-up'.

The inclusivity challenge

If the 3DP ecosystem has potential to be a transitional trigger in the same fashion as other recent technological innovations then what are the development opportunities for 3DP in the Global South? One major area for improvement is inclusivity. In regions such as Africa and India, there is still a long way to go in bringing the majority of people on-grid. For instance, 37 per cent, or 287 million, will remain illiterate in 2015 and their children will most likely follow in the footsteps of their parents and be excluded from state education, as 28 per cent of the cost of primary and secondary education has to be met by households.[4] But we argue that inclusivity can now be achieved through open source technologies and learning systems facilitated by global partnerships as 'technological change that promotes economic change, which then engenders social change, seems, to many if not most contemporary observers, to be accelerating'.[5]

In early 2012, techfortrade started examining the potential for 3DP to deliver real economic benefits in developing markets through addressing the inclusivity challenge. The grassroots UK charity, techfortrade operates in markets where there is limited manufacturing capacity and where poor (and often rural) communities are limited by their dependence on imports of technologies. Lack of infrastructure and accessibility

DOI: 10.1057/9781137365668.0008

in traditional supply chains and limited logistics means deliveries can take weeks or even months to arrive to the individuals and communities most in need.

As 3DP becomes more and more affordable and as mobile phone penetration and Internet access increase, even in the poorest countries, the possibilities for the use of 3DP becomes increasingly tangible; however, if developing countries are going to participate on an equal basis in this exciting new technology revolution and accrue many of the economic and social benefits that we believe are attainable, it is critical that consideration is given to the inclusivity of emerging industry value chains.

There are four main interventions that might address this inclusivity challenge. First, open materials. The raw materials (filaments) that are used in 3DP vary according to the purpose for which the technology is being used. The range of materials includes plastic polymers, metals, porcelain, and even wood and stone. Today, the vast majority of 3DP machines print using plastic polymers that are almost entirely sourced as virgin plastic imported mainly from China and retailing for around US$30 per kilo.

Very little innovation has taken place in the recycling of materials for 3DP, yet there are obvious potential economic and environmental benefits. A coordinated initiative involving materials science research specialists from academia and the polymer manufacturers, fast-moving consumer goods (FMCG) companies that use plastics for bottles and containers and the 3DP industry, could result in new filaments being made from readily available recycled plastics such as high-density polyethylene (HDPE) and polyethylene terephthalate (PET). In developing countries where solid waste recycling is often undertaken by informal workers earning around US$0.15 per kilo for collected and sorted plastic, recycling to create filament could create significant income opportunities whilst 'pump priming' local 3DP businesses by providing access to low-cost filaments.

Second, there are open printing resources. An increasing number of open source designs are being made available for 3DP equipment that allow low-cost filament production and low-cost printing to take place through design innovation, using parts that may be assembled from resources available in many countries. In addition, there are a growing number of 3DP products being designed bespoke to address specific needs in developing countries. Some of these designs are crowdsourced

DOI: 10.1057/9781137365668.0008

via online challenges staged on sites such as GrabCAD, an online community of over 800 thousand designers.[6]

There is currently no single design and resource library where these resources can be promoted and where product designs can be catalogued, stored, and shared. By creating and extensively promoting such a repository and by linking fledgling community 3DP initiatives to volunteer designers and expert mentors who can support the set up and running of 3DP facilities, it may be possible to accelerate the use of this technology and build local 'next generation' manufacturing skills and experience.

Third, building local capacity and capability around 3D4D is about more than just the production of fairly simple products designed to meet specific needs of poor and often remote communities. It is also about building the innovation capability and capacity to enable indigenous entrepreneurs to design and build more advanced products tailored for local markets. Just as there has been an explosion in the number of 'tech hubs' that have sprung up in Africa, Asia, and Latin America, incubating software start-ups creating applications that address local market needs; there now needs to be a focus on augmenting these hubs with collaborative makerspaces that merge the use of state-of-the-art 3DP and other complimentary electronic manufacturing equipment with software development skills.

The aim of creating makerspace initiatives is to empower local entrepreneurs with skills in using state-of-the-art technology and machinery, combined with business mentoring, to develop their product design ideas so they can create viable businesses from their ventures. In the same way as both the public and private sectors (mainly mobile and software companies) have supported the development of software tech hubs, we believe that further support should be given to create makerspaces and that this should be of interest to manufacturers of electronics and computing products aiming to build their business in low emerging markets.

Fourth, there are trials of new models for distribution. Developments in 3DP technology are gradually convincing companies involved in the physical distribution of both finished goods and of replacement components to consider alternative models that shift the physical production closer to the point of consumption. In turn, this creates an opportunity to relocate some of the value addition in the production process into the local market, thus potentially creating local economic benefits.

DOI: 10.1057/9781137365668.0008

By encouraging companies to open up their data to allow local manufacturing of spare parts from back catalogues, there is an opportunity to create a 'white market' for spare parts and components which could make the challenge of sourcing such items far less difficult than it is today. Establishing a small number of trials with companies that understand both the potential commercial and social benefits of this change to their supply chains and that are prepared to share the results of their trials, could catalyse further developments in this field.

In summary, whilst we might speculate on whether we are seeing the dawn of a third Industrial Revolution as described in the media, there is no doubt that 3DP has the potential to improve the lives and livelihoods of communities in the Global South, by providing them with the tools to establish localized distributed manufacturing capabilities and as importantly, offering the possibility to build local product innovation using next generation manufacturing techniques. By finding a use for the proliferation of waste materials as the raw material that feeds the 3D printer, we can also solve some of the problems created by waste plastic in the local environment and develop income opportunities for waste picker groups involved in local recycling.

However, building this inclusive value chain requires more than a series of individual initiatives. If this time we are serious about levelling the playing field to enable developing economies to fully exploit this new technology, we need to establish a coordinated programme of work that engages governments, universities, private sector partners across various industries, and most importantly the local innovators and entrepreneurs that can drive this transformation at a local level. This model would be very different from top-down development: it would instead be bottom-up.

Bottom-up grassroots innovation

As former Deputy Secretary-General of the UN, Louise Fréchette, reminisces: 'Strategies elaborated by outsiders may be momentarily embraced, especially if they come accompanied by big checks, but their impact is typically of short duration unless governments and societies truly "own" them'.[7] There is an alternative to the top-down development – governmental, intergovernmental, multilateral, and market-based organizations – we described earlier in this chapter, which

DOI: 10.1057/9781137365668.0008

commonly enlists policy, investment, aid and other instruments in a grand manner; namely, much smaller scale, community-owned efforts. As development scholar Adrian Smith recommended to the UN in 2011 these should focus on 'networks of activists and organizations generating novel bottom-up solutions for sustainable development'.[8] These are 'grassroots innovators'.[9] Typically deeply engaged with local people within set niches, grassroots innovators implement change through extensions of their own activities: local experiments and learning mechanisms supporting and shaping multiple, diverse projects.[10] Yet, in many cases, grassroots innovators do not have access to prerequisites to enable them to provoke meaningful, systemic transitions. These prerequisites could be overseas technology suppliers, venture capitalists, donors, suppliers, legal advisors, and so on.

In the case of knowledge and technology transfers, various 'intermediaries' play a crucial role in supporting grassroots innovators, generally in three ways: 'the aggregation of knowledge, the creation of an institutional infrastructure, and framing and coordinating local-level activities'.[11] Perhaps the most impressive example of the power of intermediaries is the one-to-one micro-finance initiative 'Kiva', which connects credit card owners to lend finance to a 'Kiva entrepreneur' in the Global South, bypassing government and other top-down institutions entirely.[12]

'For sustainability transitions ... criteria need to change sharply or else the transformation runs the risk of not being sustainable due to rebound effects and other adverse impacts'.[13] Due to the urgency of many development issues in the Global South, top-down organizations are increasingly supporting grassroots innovators through intermediaries in the Global North. NGOs and other community actors' ranks have swollen in the last few decades as a consequence in areas such as micro finance or e-governance.[14] As an example, in initiatives targeting agricultural development, private sector players gave impetus to the 'Best Bets' programme of small, like-minded innovators, achieving more than conventional types of aid to be the kind where 'research interacts directly with technology development'.[15]

Unfortunately, 'fighting poverty cannot be regarded as a matter of expanding formal markets at the expense of informality'.[16] Excitement about the apparent synthesis of indigenous innovation (such as *jugaad* in India) with top-down technology and investment strategies has seen excitement about spontaneous cultures of enterprise that might be nurtured in, and exported from, the Global South.[17] A core problem

DOI: 10.1057/9781137365668.0008

in the promotion of bottom-up grassroots innovators by top-down developers is that discourses of indigenous creativity and ingenuity – powerful tools in grassroots innovation – can be co-opted by big interests. Top-down discourses have arisen from the informal nature of employment in the Global South and the latent resource scarcity. These discourses romanticize the efforts of the poor to 'make do and mend'. In some cases, these interests elevate the relevance of incompatible niche innovations, such as Information and Communications Technologies (ICTs) or micro-finance initiatives, in order to reimagine poverty sufferers as embryonic consumers.[18] This issue arose in the case of the world's cheapest tablet (the Aakash) and the world's cheapest car (the Nano), both highly publicized, but inchoate efforts to market to the Bottom of the Pyramid (BOP) in India and push them into formal debt through micro-finance loans and the like.[19]

This trend of romanticizing indigenous innovation is carried over into top-down efforts in the Global South by elites in companies and governments to promote business-as-usual while at the same time meeting development goals – at least symbolically. India's National Innovation Council (NIC), for example, makes much of 'informal improvisation' as a method of addressing systemic issues linked to poverty as 'constraint-based opportunism'.[20] Yet, the celebration of indigenous enterprise by politicians in the Global South is cynical when elsewhere these conditions of economic and social informality lead to corruption. This is no more obvious than with intermediaries who use contractual ambiguities to become predatory brokers by siphoning funds ear-marked for local development projects.[21]

Hubs of innovation can play a premium role in providing reliable intermediaries for grassroots innovators. While major industrial centres in developing countries are obvious hubs, traditional community centres, which the BOP already make use of, including the village, are also plausible sites for investment and 'upgrading' programmes.[22] A problem here is that upgrading to innovation hubs is also a pathway to integration into GPNs that reverse or counteract sustainable development agendas, instead encouraging labour exploitation through converting local artisans into global commodity chains – a trend that has unfortunately been all too common in the Global South since colonial times.[23]

The informality of societies and economies in the Global South is also an instrumental pathway for hubs of innovation. Dense urban conurbations in megacities and swollen mid-size cities and towns link to the

DOI: 10.1057/9781137365668.0008

rural world through intensive telecommunications and subsidized transport networks that reach capacity at seasonally critical points of the year according to agriculture and other demands. Hence, areas of poverty within urban centres are also typically densely functional and hotbeds of grassroots innovation, supplying local communities with a range of services at the poverty line.[24] Development scholars David Satterthwaite and Diana Mitlin suggest a combination of local people within networks and external intermediaries is in order to negotiate and lobby local governments for formalization of services to informal settlements and for providing small funds for projects.[25]

'Having decided to import technology and the know-how – the hardware of the process of modernization – the developing country has the task of tacking the software – the recipients of technology'.[26] A consensus that bottom-up 'grassroots innovation' is a real alternative to 'market-based innovation' is growing, particularly as top-down development efforts are taking this on board as another option for policy and investment initiatives.[27] The approach is appropriate for 3D4D wherein once the 3DP ecosystem is established it can then become a self-sustaining 'circular' economy. Grassroots bottom-up change positions communities to manage their own development after intermediaries engender the initial establishment of community infrastructures through technology and knowledge transfers.

A 3D4D wishlist

'Households in the global south face entirely different challenges for "sustainability": for them, consumption may simply mean survival.'[28] So far in this chapter, we have learnt that the development status quo is primed for an intervention by 3DP. In the same fashion as other socio-technical movements, which have had a bearing in the Global South, 3DP is unlikely to be predictable in its adoption and further consequences for human development. Central to 3DP's contribution to development goals is the congruence of some or even all of the above themes in this book. The success of their harmony relies on a critical premise: 3DP could have more impact in the Global South (the developing world) than in the (developed) Global North. What remains is to bring together these key themes into a wishlist for 3D4D that is marked by the limits of the Global South.

DOI: 10.1057/9781137365668.0008

First, currently *limited markets* of the world's BOP will be a catalyst for 3DP's mass adoption. To do so, the 3DP ecosystem (technologies, materials, designs, and infrastructures) must be affordable enough to attract consumers on meagre and unpredictable incomes without drawing them into debt or contractual bondage. The market at the BOP is currently the holy grail for many multinational companies in its promise of a billion customers, which lure them into less than sincere attempts to meet the needs of those on less than a few dollars a day who in fact are largely not in a position to become consumers.[29]

Second, 3DP will assist with the BOP's desire to reuse and recycle *limited tools*. In this respect the quality of the printers and their parts are crucial to allow users to service and 'jerry-rig' themselves without accreditation or access to the Internet or international suppliers. Standard, universal parts are a must in this regard. The quality of the actual prints is an issue with 3DP. While cosmetic defects and unavoidable imperfections are less concerning for users in the Global South than in the North, prints need to be able to withstand repetitive strains and critical forces and printers must be able to work with quality materials. Moreover, designs made available for free or little cost must also be durable, efficient, and bereft of cosmetic novelties.

Third, 3DP will synchronize with existing *limited infrastructure* by supporting a local, community focus. The 3DP ecosystem will not overtax existing infrastructures or generate greater bureaucracy or demand top-down investments for longevity. Flexibility is key here and ingenuity in making the various elements of the ecosystem off-grid.

Fourth, 3DP assists learning and education despite *limited resources*. Simplicity will be vital for introducing the 3DP ecosystem into current educational regimes within restrained and underfunded circumstances. Moreover, the operation of graphical interfaces; assembly, replacement and repair; reverse engineering, formatting and touch-up of designs; and handling of materials and safety are all pressing issues. All of these areas require training and education and in turn simplicity in design and product production.

Fifth, 3DP's mass adoption would pivot off of the Global South's scope for leapfrogging due to *limited encumbrances* compared to the risk aversion and pre-existing systems found in the Global North. Hype cycles in the Global North can quickly become sapped of enthusiasm through the lobbying (malicious or not) of incumbents; however, in the Global South there are often limited or not incumbent

DOI: 10.1057/9781137365668.0008

systems in place allowing speedy transitions and massification. The 3DP ecosystem would need to meet this potentially rapid demand by being scalable.

The efficiencies of the 3DP processes go far beyond simple cost-savings on materials through careful computer-control assistance and the additive layering of objects – the orthodox picture of 3DP's benefit. If the key elements in the Goldilocks Zone of affordability, flexibility, simplicity, scalability, and quality are met within the natural course of the innovation's exponential growth, predictably following Moore's Law, then the five 'wishes' this chapter introduces above work in favour of 3D4D becoming a preeminent strategy in the context of the Global South's development in terms of its people's wellbeing and prosperity on par with the Global North.[30]

And so for 3D4D to happen effectively, there are various 'wishes' that need to be granted. As we elaborate in the following chapter, it is possible to drill down into these requirements and imagine what 3D4D will be like in the various elements of the 3DP ecosystem. Moreover, there are grassroots innovators working within and without the development nexus that we identify as indictors and even forerunners of 3D4D. Ultimately it is through global partnerships that we foresee 3D4D as being able to alleviate material poverty.

Notes

1 Reeves (2013a) '3D4D Challenge Finalist Rehearsal Day'.
2 Hulme (2010) 'Governing Global Poverty? Global Ambivalence and the Millennium Development Goals'.
3 UN (2012) 'UN System Task Team on the Post-2015 Development Agenda'.
4 UNESCO (2014) 'Education For All Global Monitoring Report'.
5 McGann and Sabatini (2011) *Global Think Tanks: Policy Networks and Governance*: 32.
6 A GrabCAD challenge staged by a US non-profit working in Nebaj, Guatemala called for designs for 3D printable children's sunglasses. One prevalent problem for the rural communities of Guatemala is Pterygium (known in the US as Surfer's Eye) an eye condition resulting from excessive exposure to ultraviolet light and aggravated by dust, smoke and other irritants. While treatment is available, resurgence is common, and preventative measures are sorely needed, since exposure to light and irritants starts at a young age. Protective sunglasses are an effective preventative tool and are

DOI: 10.1057/9781137365668.0008

especially needed for children. While relatively inexpensive adult sunglasses are readily available to the rural communities, there is a minimal supply of children's eyewear of any kind. Seventy design submissions were received of which at least six were readily usable in Nebaj.

7 Fréchette (2010) 'Foreword'.

8 Smith (2011) 'Viewpoints: "What Do You Think Should Be the Two or Three Highest Priority Political Outcomes of the United Nations Conference on Sustainable Development (Rio+20), Scheduled for Rio de Janeiro in June 2012?"': 341.

9 Smith and Seyfang (2013) 'Constructing Grassroots Innovations for Sustainability'.

10 Seyfang and Longhurst (2013) 'Desperately Seeking Niches: Grassroots Innovations and Niche Development in the Community Currency Field'.

11 Hargreaves, Hielscher, Seyfang and Smith (2013) 'Grassroots Innovations in Community Energy: The Role of Intermediaries in Niche Development': 870.

12 Schwittay (2013) 'Making Poverty into a Financial Problem: From Global Poverty Lines to kiva.org'.

13 Kemp and van Lente (2013) 'The Dual Challenge of Sustainability Transitions: Different Trajectories and Criteria': 116.

14 Madon (2009) *e-Governance for Development: A Focus on Rural India.*

15 Clark, Frost, Maudlin and Ward (2013) *Technology Development Assistance for Agriculture Putting Research into Use in Low Income Countries.*

16 Araujo (2013) 'What Have Markets Ever Done for the Poor?': 387.

17 Rangaswamy and Densmore (2013) 'Understanding Jugaad: ICTD and the Tensions of Appropriation, Innovation and Utility'.

18 Prahalad (2010) *The Fortune at the Bottom of the Pyramid: Eradicating Poverty Through Profits.*

19 Mudliar and Pal (2013) 'ICTD in the Popular Press: Media Discourse around Aakash, the "World's Cheapest Tablet"'.

20 Jeffrey and Young (2013) 'Jugād: Youth and Enterprise in India': 192.

21 Jeffrey and Dyson (2013) 'Zigzag Capitalism: Youth Entrepreneurship in the Contemporary Global South'.

22 Abrol (2004) 'Lessons from the Design of Innovation Systems for Rural Industrial Clusters in India'.

23 Heyer (2013) 'Integration into a Global Production Network: Impacts on Labour in Tiruppur's Rural Hinterlands'; Haynes (2013) 'The Making of the Hyper-Industrial City in Western India: The Transformation of Artisanal Towns into Middle-Sized Urban Centres, 1930–70'.

24 Kaufmann (2013) 'The Dharavi Paradox'.

25 Satterthwaite and Mitlin (2014) *Reducing Urban Poverty in the Global South.*

26 Atal (1972) 'Role of Values and Institutions': 89.

DOI: 10.1057/9781137365668.0008

27 Tang, Karhu and Hämäläinen (2011) 'Community Innovation in Sustainable Development: A Cross Case Study'.
28 Gibson, Farbotko, Gill, Head and Waitt (2013) *Household Sustainability: Challenges and Dilemmas in Everyday Life*: 2.
29 Karnani (2007) 'Doing Well By Doing Good – Case Study: "Fair & Lovely" Whitening Cream'.
30 See Cano (2011) 'The Cambrian Explosion of Popular 3D Printing'.

DOI: 10.1057/9781137365668.0008

5

The 3D4D Elements

Abstract: *Various elements are discernable in what Birtchnell and Hoyle describe in a framework for 3D4D. Focusing on four ideal types representing the exemplary configuration of elements for development action, this chapter scopes this framework to facilitate a better understanding of the 3D printing ecosystem. The innovation of community 3D printers, which users ideally share, is a primary candidate for a ubiquitous technology of mass adoption. In support of this technology are open source repositories of designs free from fees and legal strangleholds. Recycled materials from local sources make up the main input for 3D printing and these are made available through circular economies. Infrastructure that is independent from complex systems and energies currently inaccessible to the bottom of the pyramid is the final element this chapter considers in detail.*

Birtchnell, Thomas and William Hoyle. *3D Printing for Development in the Global South: The 3D4D Challenge.* Basingstoke: Palgrave Macmillan, 2014.
DOI: 10.1057/9781137365668.0009.

Scoping 3D4D

Chapters 1–4 gave an account of the potency for social change in the 3DP ecosystem – a combination of technologies, designs, materials, and infrastructures. Changes in the latter, in particular, radically revise how worldwide systems of global production and consumption are done. More than just another technology on the shop shelf, 3DP intimates an alternative pathway to realize what is being couched 'Development 2.0'; Chapters 5–6 go into greater detail about what a 3DP ecosystem in tune with current global development goals might aspire to be like.[1] The chapters in this part will depart from the theory of Chapters 1–4 and aim instead to bring 3D4D to life through case studies of early adopters who are implementing development now through innovative approaches to meeting everyday needs and wants in the Global South.

3D4D represents a social and technical transition that registers a possible shift both in the fortunes of the majority in the Global South, where we suggest this innovation has most brunt to bear, and in the Global North as well. Much of the media hype around 3DP concerns the latter consumers' preferred methods of procuring the 'stuff' (food, clothes, electronics, tools, novelties, shelters) they rely upon to give their everyday lives a sense of 'normality' through comfort, convenience and cleanliness: the three C's.[2] Yet, this book argues that it is in areas where the worldwide system of global production and consumption is not able to reach, at the Bottom of the Pyramid (BOP) where there are few or no consumers, that 3DP has the most brunt to bear for development.

Looking ahead, 3D4D can be viewed as involving a spate of alternative ideologies – some indigenous, some introduced – that shape the future, meeting current expectations in the Global South for a modern, global, cosmopolitan lifestyle that satisfies the three C's as much as is possible for all and not just a minority few. These ideologies certainly demand some degree of effort to adopt ways of living that might, at the moment, be unpalatable to those who set, or aspire to set, the benchmark of normal living in the Global North.

In order for 3D4D to actively invoke real change for poverty, according to established development goals, careful concern must be paid to the 'nitty gritty' of the 3DP ecosystem. To do this, Chapters 5–6 consider case studies where some, or all, of the 3DP ecosystem is being made actionable either incidentally or purposefully. By sorting through case studies

DOI: 10.1057/9781137365668.0009

of action being taken now we assess the likelihood of 3DP becoming a development enabler.

Given the constraints of 3DP there are certain configurations of the ecosystem that are most opportune for 3D4D. This configuration is not the only possibility and certain unforeseen innovations, 'black swans', could certainly create alternative pathways.[3] In Chapters 5–6, we assess the different facets of the optimal ecosystem and draw on evidence from activities happening now in the Global South. Chapter 5 teases apart different elements in the current 3DP ecosystem that bode well for the future of a 3D4D-driven transition. Chapter 6 describes case studies of these elements using examples from the 3D4D Challenge and other instances of innovation.

The first element to survey is the technology that is most conducive to 3D4D in the Global South: the printers that will be put to use on a daily basis to meet the three C's. Unlike the hype that has built up over the last few decades around home and desktop 3DP, there are many reasons this particular technology is incompatible with societies in the Global South. Instead, the community printer is elected to be the most relevant, and different scenarios around this technology are given thought.

The second element is design – chiefly, what the technologies will print out. The text and images two-dimensional printers set onto paper are part of completely different spheres of activity, distinct from the printing process, and this must be acknowledged in 3D4D. Here consideration is given to the design process and the various means available to those on low to no incomes to make things: 3D scanning, reverse engineering, open repositories, and peer-to-peer networks.

The third element is the raw resources the objects are made of, the industrial processes involved in making them suitable feedstock for 3DP and the means for transporting them safely and efficiently while limiting worldwide commodity chains unsuitable or inaccessible to the poor, as in the case of 2D printer cartridge economies. Input is not the only aspect to consider; waste outputs from production processes and the disposal of broken or discarded objects are also necessary to analyse. Optimal in this element is the potential for circular economies that fuse inputs to outputs enabling severance from worldwide systems of production and consumption.

Once all these elements are aligned there are also the various logistics of bringing them together as an ecosystem. To innovate and implement 3D4D there will also be a suite of intermediaries, investors, and local

DOI: 10.1057/9781137365668.0009

champions working together to bring this transition to fruition. The 3D4D Challenge, discussed in detail earlier is a convincing example of how disparate actors can be brought together from the public, private, philanthropic, and academic domains, in order to realize grassroots innovation. This campaign shows that a principal advantage of 3DP is 'agility'.[4] Together the elements in this chapter should provide a waystation to understanding how 3D4D is to be made a viable approach to meaningful social change. While certainly attractive to multinational corporations looking for the next stage of 'just-in time' manufacturing, agility in the production process is also a once-in-a-lifetime chance for the majority poor in the Global South currently outside of worldwide systems of production and consumption to realize desired standards of comfort, cleanliness, and convenience enjoyed in the Global North. This would be realized through consumers also becoming producers and managing their own fates.

Community printers

In our account of the Goldilocks Zone for 3DP we apprized the significance of various elements: technologies, materials, designs, and infrastructures. In the case of the element of technology, we depart from much of the current 3DP evangelism – surveyed in Chapter 2 – which points to the domestic 'home' printer as the ideal type for mass-adoption. In this section, we suggest that – at least in the Global South – a more likely candidate for a technology within the Goldilocks Zone is a community printer that users share informally or semi-formally.

Anyone who has walked the streets of a megacity in the Global South, such as Mumbai in India, might have come across the sight of office printing machines perched on the pavement with people busily scanning, faxing, and printing on them, oblivious of the other passers-by and traffic. These pedestrian print shops bear witness to the ingenuity of entrepreneurs in the informal economies of the Global South, where spontaneous solutions arise to issues difficult to conceive of in the Global North – namely, a relative dearth of domestic paper printers in homes. The informal sharing of 3D printers in semi-commercial conditions is an idea compatible with mass-adoption scenarios of a kind different to other personal devices, chiefly the mobile phone.

The community print shop is far more likely to be the vessel for proliferation in comparison to personal ownership or commercial centres, such as Internet cafes, which studies in Uganda and Tanzania show depend on a disposable income to access, due to the quality of their services and target audience of the urban middle classes.[5] A middle ground might be development projects that target the high use of public share facilities, such as pavement print shops, in order to encourage more egalitarian use and adoption of the Internet as well.[6] In the Global South, the growth of pavement print shops occurred in a much less visible way than Internet cafes, emerging from the freelance activities of entrepreneurs in the informal economy, often from the young unemployed trying to manage their meagre job prospects in the formal economy through personal promotion in the informal: printing CVs and references.[7]

The most visible element in the 3DP ecosystem is the printers: at the low end are the units on desktops with a footprint much like a paper printer; at the high end the units stand at head height and require operation with safety gear, training and industrial conditions. We can imagine a great number of technologies of the same nature, but differing in the build and specifications, through drawing parallels with other technologies. Take, for instance, the rapid spread of mobile phone handsets. The proliferation of mobiles in India is remarkable: from 8 per cent of the population to 75 per cent in five years.[8]

On the face of it, home domestic 3D printers appear to be a stalwart option for the Global South as they have many common features with mobile phones. Yet these can also be shared amongst a number of users as community printers. They are portable in both rural and urban locations; repairable, upgradeable, and hackable for little cost; they are a marker of status; and they are open to frugal innovation. This last point is perhaps the most important, as the 'missed calls' phenomenon in Africa and South Asia demonstrates, where a user deploys a catalogue of 'beeps' to initiate a conversation; frugal innovation of this sort can drive mass-adoption.[9] The innovative use of technologies should not be overlooked as core drivers of social movements and trends are never straightforward. The most obvious candidate is the RepRap, introduced in Chapter 2, and its many variations. As inventor Bowyer speculates, once community RepRaps are established, frugal innovation in the self-replication of printer parts could drive the adoption of 3DP in the Global South.

Alternatively, the mobile phone could be the very technology that initiates 3DP. For instance, Motorola and 3D Systems propose to release an 'open hardware' handset frame, dubbed Project Ara, which would

DOI: 10.1057/9781137365668.0009

have an 'endoskeleton' frame compatible with modular 3D printed parts, customized to the owners' specifications.[10] This type of innovation is of particular importance in the Global South, where different cultural factors such as fashion can play a role in mass-adoption. For instance, a survey of Indian phone users found they value being able to personalize handsets' ringtones, wallpapers and covers to increase their personal relevance.[11]

A possible tipping point for 3DP is also possible in an entirely unpredictable combination of elements, old and new, arising from outside of the technology itself, as Historian David Nye articulates: 'any trend that seems obvious, and any pattern that seems persistent, may be destabilized by changes in the economy, changes in technology, or some combination of social and technical factors'.[12] In Africa and South Asia, a pivotal influence in the spread of mobile handsets has been informal unlocking and modification services that give access to the most cost-effective prepaid plans and subvert handset manufacturers' inbuilt restraints.

We have thus laid out some core characteristics of the optimal technology element, community 3D printers, which will provide a framework for our interpretation of the case studies in the next chapters. We consider now the next element in our ecosystem that underpins the Goldilocks Zone 3D printer for the Global South: open repositories of designs.

Open repositories

There is not much use in a 3D printer without designs to print – in the same fashion as a paper printer requires a computer, word processor and text or images, in order to function fully. Beyond the technical points of the technology, much of the novelty of the 3DP ecosystem comes from the community interaction and sharing of ideas and useful applications online. Indeed, the 'prosumer' idea considered in Chapter 3 is that the consumer becomes responsible for the nature of the objects they use. Web 2.0 – that is, open source, user-interactive, online social network platforms – enhance this process of consumers becoming producers, or, taking this one step further, 'produsers'. Community decision-making, continuing leadership turnover, gradual quality development, and individual rewards from common property are all part of the produser idea.[13] Curiously for us, all of these features are extant in many of the open repositories for 3DP designs now available online.

DOI: 10.1057/9781137365668.0009

In the Global South, the cost of software is prohibitive and a major deterrent to the wider promotion of literacy, numeracy, and employment opportunities to those on the poverty line. The significant cohorts of users unable to afford individual or organizational software licences in the Global South were a driving force in the open source software movement in the first place. There are now operating systems (Linux), word processors (LibreOffice), browsers (Firefox), and even specialist programmes for audio (Audacity) and 3D design suites (Sketchup). The importance of open source cannot be underestimated. A comparison of license fees to a country's Gross Domestic Product (GDP) per capita (average individual income) shows that this is far higher in developing countries. For example, a comparison of Microsoft Windows XP license costs as a percentage of GDP shows 0.19 in the US, 0.32 in the EU, and 0.48 in Oceania: 19.19 in Bangladesh, 24.12 in Cambodia, 26.19 in Central African Republic, and a staggering 70.96 in Ethiopia.[14]

The major developers of 3D printers are well aware of the need for open repositories of designs that are useful for everyday life objects and not just for novelties. One of the main suppliers of domestic home 3D printers, Makerbot Industries, set up Thingiverse in November 2008 as a companion to the Replicator model printer. Now stocking more than 100,000 designs the site is entirely open source with categories of collections, including the keyword 'Household': 'utilitarian prints that make everyday life easier'. The files are standard stereolithography (STL) format and generally downloadable on bandwidths of Internet access found in the Global South.

Aside from cost, there is also the issue of the graphic interface that needs to take into account the education standards of all users in the Global South. The ideal open repository is one that incorporates templates of common designs with image or picture representations of the objects. Such features as touch-screens with image templates are within the realm of current possibility and have seen success in rural communities in India.[15]

There are also less well-known online repositories that emphasize mass-customization through user interaction and collaboration, such as Israeli start-up, ShapeDo. A community-driven platform, ShapeDo users work together to create objects through an online interface that displays a graphic of the object in real-time. The site is free to use with open source licenses and requires no additional coding skills to manipulate the object designs.

DOI: 10.1057/9781137365668.0009

It is not inconceivable for there to be repositories in the Global South that function like barter economies, wherein those solutions that evolve to be most appropriate for local conditions are captured in databases and sorted using keywords and other methods that can then be shared and swapped depending on the situations and resources of the users. The task for grassroots innovators and their intermediaries is to actively participate in the facilitation and testing of pre-existing objects, some presumably with long histories of use. A prime service will be to reverse engineer objects arising from indigenous solutions using 3D scanning, intuitive catalogues, and design principles to make the prints function effectively with the material and structural restraints of the 3D printers available on the ground. A simple possibility is an open repository allowing the resizing of objects without needing to change the Computer Aided Design (CAD) file. Catalogues of 3D scans for 3DP are already under development for archival and historical purposes. The idea here is that replicas could be manufactured from high resolution scans in the event of decay, loss, damage or neglect.

Open repositories need not only stock designs to print, but also, as in the case of the RepRap, parts for printers themselves bundled with software.[16] Likewise, there is no reason innovative solutions to specific issues within communities in the Global South cannot be 3D scanned locally and converted (with some additional processing to make scaffolding and material compensation). This is the vision of affordable consumer 3D scanners. Makerbot's Digitizer is one such example that offers high resolution scanning and simple conversion to 3DP file formats, although it cannot scan everything: shiny, reflective or 'fuzzy' objects are as yet incompatible. The potential of 3D scanning in the Global South is perhaps more important than online open repositories, as scans can be shared locally for issues that come up in communities facing similar problems.

Recycled materials

It is all fine and good to locate a 3D printer in the Global South in a community-run facility with open repositories offering catalogues of objects and parts useful for those surviving on the lowest of incomes. However, in order to service these demands there must also be stocks of materials available to print, just as a 2D printer requires a ream of paper and cartridges of ink powders. By introducing this element of feedstock, it immediately becomes clear that what this entails is a further tether to global supply chains of the sort reviewed in Chapter 2, both for procurement of materials in printer-compatible formats and for disposal of waste products deriving from the printing process (scaffolds, defects, residues). As Chapter 3 made clear, objects acceptable for consumers in the Global North and the minority of high and middle income earners in the Global South – already accustomed to regularly replacing parts and accessories – are an unacceptable financial burden for the BOP. In this section, we explore some models for providing feedstock for 3DP in the Global South without adding debt commitments and other pecuniary constraints on the poor.

We especially emphasize the continuation of existing systems for 3D4D. Informal and formal systems already operating within societies in the Global South are the best candidates to provide feedstock for the user-led circular economies introduced in Chapter 3. A major area of interest for us here is the practice of waste-picking. Poorly paid, dangerous, demeaning, unskilled, and unpredictable, this practice is an income base for the vulnerable sections of the developing countries. Paradoxically, but understandably, those whose incomes rely on this practice are the most vocal for access to more waste. The Global Alliance of Waste Pickers' mission statement rejects incineration and landfill-based technologies because they restrict access to informal sorting and collecting and make waste sites more dangerous for itinerant gatherers.

Waste-picking is not simply individuals sorting through landfill; the practice involves complex social hierarchies, demographics and, indeed, pecking orders. In Nigeria, for instance, there are four distinct levels. First, the most vulnerable: children who pay for their education through foraging for waste after school and on weekends. Second, there are the wandering door-to-door collectors of waste with their own territories and established customers. Third, there are those with access to communal small-scale waste sites. Finally, there are those able to gain access to the large open-air landfills, valuable sites for the recovery of recyclables.[17] As this list demonstrates, waste-picking is rife with the exploitation of the vulnerable. Indeed, some scholars argue that the growth of waste-picking owes much to the structure of supply chain capitalism and the globalization of production networks, as consumer goods 'are not only manufactured but increasingly recycled in poor countries'.[18] In India, the structures in waste-picking appear even more

DOI: 10.1057/9781137365668.0009

complex as there are also traditional 'fine-grain' caste considerations within larger hierarchies.

Development scholar Yujiro Hayami and colleagues show that in Delhi, India, waste-pickers are drawn from the ranks of migrants to the state, living in slums. There are two groups *kacharawala* (waste-pickers) and *kabadis* (waste-collectors) at the bottom of India's social hierarchy. Both small self-employed informal agents, pickers need no capital for picking up public waste – paper, plastic bottles, and aluminium cans – while collectors buy their waste from the producers (households and small business) in cash, for which they must have some existing capital.[19] Collectors are also able to realize economies of scale through the use of carts, while pickers simply carry waste on their backs. In considering development interventions, pickers are more prone to chronic poverty as collectors are able to be more socially mobile and can become dealers or even wholesalers of waste.

The introduction of standards and technologies to landfill sites is problematic as shown in the wake of the 2010 Oscar-nominated documentary *Wasteland*, which follows US artist Vik Muniz's project to convert waste gathered from Rio De Janeiro's *Jardim Gramacho* into expensive artworks for the international market, sold for US$300,000 and donated back to the waste-pickers. In 2012, the open-air landfill was closed and waste diverted to *Seropedica* – a waste-treatment centre that converts the waste into energy, but excludes waste-pickers.

Of the 15 million people worldwide who make waste-picking their livelihood, the majority are in the Global South, often in places where municipal, formal systems are not allocated for waste removal and recycling.[20] Moreover, these regions are willing to accept e-waste (computers and other electronics) that are unacceptable for handling and processing in the Global North – this is often imported illegally and buried, burnt in the open air or dumped into surface water bodies.[21] MICs take on e-waste imports as they can reuse the raw materials found in e-waste streams.[22]

The formalization of waste-picking is challenging and requires a deep sensitivity to the various unintended ramifications that could spin off the top-down interventions critiqued earlier in Chapter 4. An example of this is Buenos Aires' *cartoneros* and the Programme for Urban Recuperators, launched in 2003 by the municipal government.[23] The *cartoneros*, like many waste-pickers, informally collect a range of materials, including aluminium, copper, and all types of plastic – materials

DOI: 10.1057/9781137365668.0009

useful for 3DP. However, research shows that efforts to formalize their labour, while increasing social legitimacy and decreasing stigmatization, is only likely to benefit those able to ally themselves with cooperatives granted municipal contracts. This sort of arrangement could have adverse effects – interpersonal conflicts, longer commuting times, and off-putting scheduling. This example shadows other similar top-down intervention attempts around the world. We argue that in the case of 3D4D, bottom-up 'grassroots' efforts are a more viable alternative for meaningful change.

Grassroots efforts could target the short term by encouraging and supplying protective equipment to waste-pickers (invariably children), such as gloves, footwear, and tools to sort waste, as well as vaccination against tetanus.[24] In the early 1990s, the non-governmental organization (NGO) Exnora International sponsored the formation of neighbourhood associations to manage waste collection in Chennai, India. The community-based groups also encouraged waste to be perceived as a resource and introduced recycling methods to their services.[25] Also, local initiatives enlisting open source knowledge and technology transfers are promising low-cost possibilities for 3D4D. For instance, the website *africagadget.com* which provides a plastic recycling press that transforms plastic waste for local communities in Africa

The profit from 3DP for waste-pickers in the Global South is obvious in the case of Kodjo Afate Gnikou, from Togo, West Africa, who put together the first 3D printer made from e-waste he had gathered (rails and belts from old scanners), as well as new parts, in all costing US$100 to build.[26] He uses the printer to make objects that are useful around the home, but difficult to procure in Africa. Gnikou inspired the W.AFATE project by the Woelab Community to start a crowd funding venture to publicize the architecture of a printer made from IT debris in 10 easy-to-follow steps.

3DP infrastructure

> John Dimatos, the former Head of Applications for Makerbot works with UNICEF as an innovation specialist... One of the major comparisons that comes up with 3D printing is the meteoric rise of cell phones. Dimatos points however that, unlike cellphones, 3D printers can't be charged up in town and used later back home. Solar cells aren't up to the task of powering these very wattage-hungry machines either.[27]

DOI: 10.1057/9781137365668.0009

The infrastructure in place for 3DP to become significant for development is a final element that needs consideration. As Dimatos above notes, reliable and mains power is a vital issue for 3D printers with both features being in short supply in the Global South. To be sure, energy is a logistical issue although conceptually community printers can be run on solar power. Conventional versions of 3DP technologies are available using rechargeable batteries, which allow them to function outdoors or in a power cut.[28]

For some purposes, however, it is desirable to take a broader view of the subject of energy in the Global South. At a community-level solar-power technologies are certainly promising, for example, the Sarvajal organization launched a network of 'water ATMs' serving 110,000 rural customers in India.[29] In order to be financially viable community technologies need a degree of scale – in this case, 800 franchisees across the county – to make a profit. On a less savoury note, energy theft is rampant in the Global South where authorities have limits to their resources for dealing with illegal tampering and connections. In some parts of India's cities, this adds up to a form of subsidization for domestic and light commercial use. Yet, there is also the possibility of running networks of community-based organizations (CBOs) that buy and sell sustainable energy and punish pilferage, as with hydropower in Nepal.[30]

3DP could become ubiquitous regardless of reform through access to energy supplies that are not currently well documented. The revealing, 2013 documentary, *Katiyabaaz* (Powerless), shows the ingenuity of energy intermediaries in the Global South who routinely modify and customize the decrepit energy infrastructure so their 'clients' can access power illegally. Business-as-usual – that is, energy theft – could well continue unchecked, as there is little hope for formalization of the grid in the Global South. 3D4D might, in this case, occur under the radar.

Some of the more outlandish ideas for 3DP infrastructure are still a long way off, although not inconceivable: printing houses or even entire suburbs from scratch. One indicator of 3DP's future applications in the Global South is projects to additively manufacture lunar buildings. As far-flung as this seems, the Moon is a costly location to build in, thus there are resource and energy limitations due to this austere habitat. Scientist Silvia Benvenuti and colleagues consider that the first major benefit of 3DP in resource-constrained conditions is the ability to manufacture complex geometries using 'D-Shape' technology (a gigantic plotter capable of printing in metals and concrete) to make buildings to

DOI: 10.1057/9781137365668.0009

larger scales than usual printers.[31] A 'printing rover' would be a mobile manufacturing unit, laying down infrastructure intelligently. Most relevant for 3D4D are plans to use *in situ* resources, sands and soils. To limit transport costs, the '3D Printed Building Blocks Using Lunar Soil', funded by the European Space Agency, prototypes infrastructure made from lunar regolith and a novel direct manufacturing (3D printing) technique.

In this vein a pathway for grassroots innovations in 3DP infrastructure is the utilization of local materials onsite or near to hand. In 2011, the Industrial Designer Markus Kayser built a solar-powered 3D printer that manufactures glass objects from sand collected from the surrounding environment (in the case of his experiment, the deserts of Siwa, Egypt).[32] Although somewhat portable (Markus is shown hauling the solar sinter by foot), the scale of this printer means it is not conducive to the home printer market, but certainly could be a viable community 3D printer. Printing suburbs for the poor using solar and sand might bring down costs currently leading to inadequate and ailing infrastructure in the slums of megacities.[33]

The inclusive grassroots innovation showcased in Chapter 4 has its apparent downside in purportedly lacking the capacity to address disconnects at the political, economic, and infrastructural interface that underlie unsustainability: these are the 'big problems' of development. A method of bypassing top-down bias in the implementing of social change is through the recapture of infrastructure management from central control. Campaigns mobilizing 3DP for infrastructure patching and caretaking at the community scale could be realizable through the conversion of locally sourced materials into printable feedstock for roads, structures, and shelters. While current infrastructure work requires specialist expertise and equipment, 3DP's capacity for agility and customization are key elements here. An inkling of this is already being imagined in applications of 3DP to disaster relief, when the process becomes faster, cheaper, and more versatile. According to Steve Haines, mobilization director for Save the Children: '3D printing could make a huge difference to emergency responses, saving a fortune by printing things like tools, basic items and equipment on the ground from recycled materials, rather than flying them in from other countries'.[34]

In this chapter, the skeleton of a system for 3D4D to be deployed effectively in the Global South was assembled according to the various parts of the 3DP ecosystem now perceivable, either partly or fully. In the

DOI: 10.1057/9781137365668.0009

next chapter, we aim to bring these elements to life through drawing on case studies of grassroots innovators working with some or all of these elements now. Some of these case studies are drawn from the 3D4D Challenge 2013, summarized in Chapter 6. Others are part of networks to these innovators or other NGOs and community initiatives.

Notes

1 Brainard (2009) *Global Development 2.0: Can Philanthropists, the Public, and the Poor Make Poverty History?*.
2 Shove, Pantzar and Watson (2012) *The Dynamics of Social Practice: Everyday Life and How it Changes.*
3 Taleb (2008) *The Black Swan: The Impact of the Highly Improbable.*
4 Vinodh, Sundararaj, Devadasan, Kuttalingam and Rajanayagam (2009) 'Agility through Rapid Prototyping Technology in a Manufacturing Environment Using a 3D Printer'.
5 Mwesige (2004) 'Cyber Elites: A Survey of Internet Café Users in Uganda'; Mercer (2006) 'Telecentres and Transformations: Modernizing Tanzania Through the Internet'.
6 Haseloff (2005) 'Cybercafés and Their Potential as Community Development Tools in India'.
7 Jeffrey (2010) *Timepass: Youth, Class, and the Politics of Waiting in India.*
8 Agar (2013) *Constant Touch: A Global History of the Mobile Phone.*
9 Donner (2007) 'The Rules of Beeping: Exchanging Messages Via Intentional "Missed Calls" on Mobile Phones'.
10 Velazco (2013) 'Motorola Taps 3D Systems To Produce Parts For Its Crazy Modular Smartphones'.
11 Venta, Isomursu, Ahtinen and Ramiah (2008) '"My Phone is a Part of My Soul" – How People Bond with Their Mobile Phones'.
12 Nye (2007) *Technology Matters: Questions to Live With*: 32.
13 Bruns and Schmidt (2011) 'Produsage: A Closer Look at Continuing Developments'.
14 Ghosh (2003) 'Licence Fees and GDP Per Capita: The Case for Open Source in Developing Countries'.
15 Rao (2003) 'Information Systems in Rural Indian Communities'.
16 Pearce (2012) 'The Case for Open Source Appropriate Technology'
17 Adama (2012) 'Urban Livelihoods and Social Networks: Emerging Relations in Informal Recycling in Kaduna, Nigeria'.
18 Dauvergne and LeBaron (2012) 'The Social Cost of Environmental Solutions': 415.

DOI: 10.1057/9781137365668.0009

19 Hayami, Dikshit and Mishra (2006) 'Waste pickers and Collectors in Delhi: Poverty and Environment in an Urban Informal Sector'.

20 UNEP (2012) 'Vital Waste Graphics 3'.

21 Nnorom and Osibanjo (2008) 'Overview of Electronic Waste (E-Waste) Management Practices and Legislations, and Their Poor Applications in the Developing Countries'.

22 Widmer, Oswald-Krapf, Sinha-Khetriwal, Schnellmann and Böni (2005) 'Global Perspectives on E-Waste'.

23 Chronopoulos (2006) 'The Cartoneros of Buenos Aires, 2001–2005': 1.

24 Hunt (2013) 'Formalization Beckons: A Baseline of Informal Recycling Work in Buenos Aires'.

25 Anand (2000) 'Co-operation and the Urban Environment: An Exploration'.

26 Singh (2013) 'West African Inventor Makes a $100 3D Printer From e-Waste' .

27 Guess (2013) '3D Printing for Africa's Development' .

28 Nathan (2011) 'Solar-Powered RepRap Prints Even When the Power is Out'.

29 Hu (2013) 'Evolving Paradigms of Manufacturing: From Mass Production to Mass Customization and Personalization'.

30 Sovacool, Dhakal, Gippner and Jain Bambawale (2013) 'Peeling the Energy Pickle: Expert Perceptions on Overcoming Nepal's Electricity Crisis'.

31 Benvenuti, Ceccanti and De Kestelier (2013) 'Living on the Moon: Topological Optimization of a 3D-Printed Lunar Shelter'.

32 Jobson (2011) 'Markus Kayser Builds a Solar-Powered 3D Printer that Prints Glass from Sand and a Sun-Powered Cutter' .

33 Nasman (2013) 'Dutch Architects Say 3D Print Technology Could Solve the World's Housing Problems'.

34 Koebler (2013) 'Is 3D Printing the Future of Disaster Relief?'.

DOI: 10.1057/9781137365668.0009

6
3D4D Indicators and Forerunners

Abstract: *In this chapter, case studies are given of various actors relevant to the technologies, designs, materials, and infrastructures of 3D printing. These actors are dubbed in this book 'indicators' and 'forerunners' because they give a guide to how this niche innovation might scale up to become socially significant. From a 3D printing 'go-to-guy' to an entrepreneurial philanthropic venture to turn stone powder into jewellery, the case studies in this chapter draw on insights from research in the Global South. The forerunners include iLab//Haiti, a project to encourage a community 3D printer, and the Ethical Filament Foundation, a project to turn landfill into plastic wire filament.*

Birtchnell, Thomas and William Hoyle. *3D Printing for Development in the Global South: The 3D4D Challenge.* Basingstoke: Palgrave Macmillan, 2014.
DOI: 10.1057/9781137365668.0010.

 DOI: 10.1057/9781137365668.0010

The 3DP 'go-to-guy'

The following sections in this penultimate chapter provide case studies of the elements in the 3DP ecosystem in order to assess 3D4D's scope for scale in technologies, materials, designs, and infrastructures. The first of these case studies is what we term a 3D4D indicator. In many countries in the Global South, it is a challenge to source 3DP. The concept of intermediaries, introduced in Chapter 3, is apt here in considering the procurement of the first element in the 3DP ecosystem: the technology of the printer itself. The emergence of local suppliers in India represents the emergence of intermediaries and 3DP 'go-to' people.

All sorts of structural impediments are pitted against early adopters in the Global South: customs and import restrictions, engrained corruption, political embargoes against overseas companies, and various local political and legislative inconveniences. Grassroots innovators in these regions have been drawn to the open source RepRap and the now well-established online networks that include hosts of worldwide amateurs and experts all sharing their know-how. In India, the large surplus of tertiary engineering graduates face unemployment and post graduation career gaps leading to difficulties in consolidating work-experience with their qualifications.[1] Many choose to 'abjure the realm' following established circuits to skilled migration, others motivate themselves to launch start-ups and entrepreneurial ventures, often in constrained conditions with limited resources. These individuals' activities can in many cases overlap with the interests of bottom-up grassroots innovators because they set in place complimentary pathways and routines with governments and firms in areas of common ground as suppliers of equipment, knowledge, finances, and so on.

Online, open source networks of developers worldwide prove to be a boon for aspirational entrepreneurs and enthusiasts for 3DP in the Global South. In January 2012, the *Sunday Times of India* hosted a feature article on a local 3DP enthusiast: Mumbai-based 'engineering whiz', 22-year-old Karan Chaphekar, who had built the 'first low-cost 3D printer in India'.[2] Urbane and modest, Karan is waiting in the foyer of the hotel on a bustling lane-way in the Mumbai suburb of Fort. We dodge traffic across the road to a place where we can get refreshments and talk, past the crowds of young men perching on motorcycles gazing at open-air television screens outside tea shops – there is a decisive cricket game being telecast live. Immediately after signalling to the waiter to quell the

DOI: 10.1057/9781137365668.0010

tinny music, he launches into a survey of the 3DP 'scene' in Mumbai and India more generally. It turns out the mega-city is a divining stick for emerging technology trends in the subcontinent.

For Karan and other engineering students in India's ultra-competitive education market, the skill of frugal innovation is a distinct advantage in their fixed paths in order to differentiate those students with genuine interest in learning (for instance, in the art and science of robotics), from those merely chasing a college testamur. In the late 1990s, a generation of graduates forged international circuits to the innovation hubs of the US, notably Silicon Valley, off the back of skills in antiquated software. In order to stave off the threat of a systemic date-time oversight – the Millennium 'Bug' – in major firms' Information and Communications Technology (ICT) servers, cohorts of low-cost graduates were 'parachuted' in to perform the menial but vital task of implementing remedial software patches.[3] Ever since, India's education system has sought to keep abreast of cost-effective and do it yourself (DIY) innovations in ICTs, with those graduates able to command software and hardware flexibly going on to foster start-ups and ventures.

The 3DP movement in India is still nascent. Karan gained inspiration from a local innovator, his neighbourhood mentor in Thane, whose work in the field of medical software exposed him to the potential of 3DP. The mentor began experimenting with the technology and from there became a supplier of printers and parts to his client networks who rebrand them for local resale. Together their network spans to other states in India, Jodhpur in the North and to pockets of enthusiasts in South India.

Karan talks about the 'dream' of 3DP – to make objects locally rather than have to ship them. The key thing here is to find new niches in national manufacturing – many companies in India make mobile phones for the domestic market. As in the US and UK 3DP offers small innovators a unique suite of new possibilities to shape objects for local needs. Karan enthuses that this is the reason for a very vibrant community of open source 'RepRap' developers in India.

Karan is hawkish when it comes to government support for 3D4D. For Karan 3DP has the potential to give those on low incomes in India a measure of normalcy and cosmopolitanism through the provision of objects that are both useful for everyday survival as well as for religion, celebration, and wellbeing. Karan opines that the government should support the 3DP market as much as it does for ICTs despite its nascence.

DOI: 10.1057/9781137365668.0010

He gives the example of the community printing of small-runs of custom spiritual icons (a practice that is incredibly diverse across India's religious landscape); these could be recycled after community events, thus impacting upon household incomes indirectly at the Bottom of the Pyramid (BOP) who compulsorily purchase these icons new for single events in a cyclic fashion. Novelties and 'knick-knacks' also contribute to people's sense of participation in society, no matter how low their incomes. Karan is undecided about whether 3DP will fix poverty in India; however, he is adamant that it will allow India's cohorts of unemployed graduates to create opportunities for themselves within the region without needing to chase circuits overseas in order to gain capital and proof-of-concept. He envisages this having a knock-on effect for local informal markets. After three years of building and selling 3D printers and parts based on the open source RepRap, Karan set up KCbots in 2013 with UK designer Oliver Blackwell, and the Kube and Kube Mini 3D printers are competitively priced for the Indian market and made in Thane, India.

Remarkably, Karan is not the only young grassroots innovator taking advantage of spontaneous online communities to pioneer 3DP in India. On 2 February 2014, an interview conducted by *The Times of India* with 15-year-old Mumbai resident, Angad Daryani, documents a similar story as Karan's: 'DIYer' Angad built his own 3D printer after his father refused to buy him one. He found that it only cost half the retail price to make one from scratch, under US$320, 'a price he claims is the cheapest in the country'.[4] Like Karan, Angad also aspires to be a 3DP entrepreneur through assembling printers and distributing them to the local market. These examples appear to defy received wisdom, showing that bottom-up innovators and their intermediaries enable communities at the BOP in the Global South to participate in the world economy, in the process serving local markets through informal and community networks.[5]

Turning grit into gold

In 2012, a group of Australian undergraduate students in engineering, design, management, and social sciences found themselves in the destitute outer suburbs of one of South India's most rapidly growing ICT hubs, Bengaluru. Their task was simple: find a way for the wives and children of local quarry-workers, scratching a living from extracting stone for the city's booming building sector to empower themselves through an

DOI: 10.1057/9781137365668.0010

entrepreneurial venture that did not require any changes to their current condition. This was no mean feat. Their mission was proving far more difficult than it had first seemed for this intrepid group of aspirational entrepreneurs simply because there is precious little left to spare in Indian society. Tier after tier of caste-bound occupations, each with their own niche areas of expertise and resources to make a living with, make up a complex hierarchy from the Brahmin priests who perform recitals at weddings to the scheduled castes who clean the streets. The sheer efficiency of this society spurred Anthropologist Louis Dumont to call his book on caste *Homo Hierarchicus*.[6]

The task came about through a unique institution, the 40K Foundation, which gives students a taste of aid-work through immersing them in the Global South on short but rich placements. The initiative's core innovation is the realization that apart from sourcing funds for charities through solicitation – a practice now common on the high streets of major cities in the Global North worldwide – university students are also able to offer their educational skills in teaching by gaining experience and imparting their knowledge at the same time (an alternative to going on holiday). They do this via 'pods': tiny outposts with a locally trained teacher and a suite of touch screens sporting step-by-step software, which students from any background are able to quickly master.

We have already seen so far in this chapter how latent intermediaries exist in the Global South who are able to provide the technologies for 3DP to the wider population once grassroots innovators get to work. Yet what about the materials for these printers to produce objects? These are currently part of global supply chains relatively inaccessible to the poor in the Global South. This is because of cartridge economies. Stocks fed into paper printers typically come in the form of branded cartridges, notoriously lacking in cross-compatibility, which allow the printer companies to control the price and amount of ink powders within them. This practice of artificially producing cartridge economies is already taking place with 3DP, which in some cases use identical cartridges to paper printers for safe distribution of the material powders. Yet, the open source 'RepRap' printers discussed earlier in this book do not rely on branded cartridges; instead, these use stock standard thermo-plastics available in reels of filament from many hardware outlets. In the Global South, this sort of feedstock is far more appropriate and we observe how grassroots innovators are working with this format later in the chapter.

DOI: 10.1057/9781137365668.0010

There are many locally sourced materials that could feasibly be fed into printers in the future as powders (a format more conducive to finer and stronger prints). These could become the basis of micro-ventures and bolster incomes at the BOP. The print-to-order company, Shapeways, already offers objects from powdered plastic, resin, metal, and even stone.

The initiative described at the beginning of this section is useful for this book because of the solution to the dilemma of the quarry-workers' wives that one of the students came up with in Bengaluru. While even the small chips of stone that fell onto the floor of the quarry were gathered up to be used as bitumen for roads, there was one thing these labourers had in droves and could find no use for: stone powder. An honours-year engineering student looking for a thesis project, Kimberley Abbott, had spent her time with the 40K Foundation pondering the problem of the quarry-workers' wives. Walking around the quarry with her colleagues, she recognized an opportunity compatible with her interest in materials science. Sitting with the community's women she saw how skilful they were at converting waste-pickings into necklaces and bracelets once the designs were in place for them to follow. Kimberley saw her opportunity: could the powder lying about the quarry floor also be turned into jewellery?

The logistics of this grassroots enterprise were not as challenging as Kimberley had supposed. Following some lightning site inspections in a Bengaluru autorickshaw, she was able to put together a prototype of the business model she had in mind. 'Roka' (word playing off the source of the material) would be a boutique supplier of limited edition jewellery, hand-crafted by the quarry women, with profits being returned to them. Roka is a social enterprise to use the economic empowerment of women and education to break the cycle of poverty in India. So far the charity initiative has been a success with respectable sales back in Australia and online to the international market.

The aptness of Roka for 3D4D should be obvious: the use of waste products in the Global South for outlets in the creative economies in the Global North offers a pathway into realizing bottom-up change without the intermediaries, agents, and institutions being co-opted in the process.[7] While not in a position to scale up to compete with mainstream jewellery producers, Roka instead uses its hand-crafting roots in Indian quarry labour as its raison d'être. The production of simple, charitable products is a key component of the marketing of Roka. Each piece is unique and traceable back to the individual who made it.

DOI: 10.1057/9781137365668.0010

If 3DP is to have an impact in the Global South, then materials will need to be sourced that are ethical, sustainable, and not liable to promote further exploitation. Already in this chapter, we have surveyed the 3DP indicator for sourcing the technologies and in this section, materials. What remains in the 3DP ecosystem are the designs and infrastructures: crucial elements with what appear to be insurmountable hurdles.

Demystifying design

According to Lisa Harouni, Co-Founder and CEO of the design company Digital Forming in London, one of the big misconceptions about 3DP is that everyone is going to have to learn how to use professional design programmes.[8] In fact, what her company and others are innovating are online websites that offer 'templates' of objects that can be customized or 'tweaked', perhaps in future by haptic controllers that allow a 'feel' of the object. What Digital Forming and other companies offer to their clients is an interface to manipulate libraries of designs, which can be changed with no prior coding or Computer Aided Design (CAD) experience. As 3DP allows one-offs, there is much scope for customization to local conditions. As with the case of the 3DP 'go-to-guy' in India, Karan Chaphekar, the intermediaries who play core roles in 3D4D need neither necessarily be philanthropic nor politically motivated. Another case in point is in the development of software solutions that allow people with design ideas or 3D scans to create them without purchasing licenses for CAD software and learning complex coding. There has been much work in recent years on user interfaces for 3D design software that require no prerequisite skills or training by large software companies, including the recent Google Sketchup, and also many small players. Some of these also include repositories: Sketchup provides 3D Warehouse, which has accumulated what one commentator calls a 'visual dictionary' of 3D designs.[9]

Jayesh Salvi founded 3DTin, the world's first 3D modelling tool that runs in a web browser, in 2010 from his native Mumbai, India. Via videoconference from Canada, where he is currently working, he explains the reasoning behind his innovation. Relocating back to India in 2009 after earning his degree in Computer Engineering at the University of Minnesota, he sought to satisfy local demand for an online tool that would allow users in India who might not have access to powerful

DOI: 10.1057/9781137365668.0010

computers to run CAD algorithms in the cloud even on a tablet or simple laptop. Hence, 3DTin's motto is '3D modelling for everyone'. Committed to 3DP, 3DTin easily exports designs to 3D printers and this is a feature Jayesh ensures complies with both open source and commercial units.

Jayesh is adamant that what tools like 3DTin allow is for 3DP to become accessible to the vast majority of people – it democratizes design. Because Jayesh developed the design tool in the Global South, he was sympathetic to the conditions in this part of the world and the extant limitations in access to resources. Part of the success of 3DTin is the use of pre-made templates that can be combined, warped, and otherwise, manipulated visually. This software tool is one step in a process of evolution towards the use of haptic controllers, which allow the user to 'feel' designs and alter realistic 3D images via touch. The use of stock designs that can serve as guides for users is also being trialled in repositories for actual 3D scans rather than just design files.

Progress in the democratization of design is also the agenda of many not-for-profit, public and educational institutions that seek to make 3D graphic interfaces more accessible and affordable. Their goal is to archive material objects for posterity and preservation. This movement involves the digitization of collections in museums, galleries, and universities around the world. There is much overlap between these efforts and those in the Global South to collate indigenous and ad hoc objects in order to create repositories of universal items.

Digital 3D repositories make available collections of 3D scans from archives, which can be downloaded and 3D printed: fossils, tools, ancient weapons, insects, flora, fauna, jewellery, and even scaled buildings. Jayesh's efforts to develop an affordable and intuitive interface to create designs for 3DP is just one part of the democratization of design to make 3DP fit-for-purpose in the Global South. Efforts to 3D scan all manner of objects for posterity are conducive to 3D4D as it promotes catchments of objects for frugal innovation that are printable on demand rather than shipped and stored in inventories. What can be imagined here are online libraries of designs in special centres, which are fully searchable through keywords and metadata based on the needs of the individual or the wider community. Once the digital file is located and downloaded, it can then be manipulated in software like 3DTin to make it compatible with local conditions, as well as desirable to the user (with the addition of colour, decals, symbols, and so on). The materials for such ethical print shops could be sourced from subsidized government staples (petrol, corn) and

DOI: 10.1057/9781137365668.0010

automated for efficiency and reuse of materials where possible through shredders and atomizers.

This chapter has so far undertaken three snapshots of 3D4D highlighting observable progress being made today in currently unrelated areas of the 3DP ecosystem – technology, designs, materials, and infrastructures. As Chapter 3 articulated earlier, bottom-up development includes those in need in the processes of transition through interventions that go beyond aid, setting in place *sustainable communities.*

First, there are the technology intermediaries such as Karan Chaphekar who through their entrepreneurial pioneering, and niche interests partisan to the status quo, provide technical skill and access to resources 'on the ground' in the Global South. While only incidental to development goals, intermediaries buttress efficient and inexpensive knowledge and technology transfers. Intermediaries are vital links in the grassroots development chain, laying down supply chains, lobbying for local infrastructure and policy-support and running in parallel to development agendas as exemplified by the efforts of Karan to invoke a 3D printer suitable for conditions in the Global South creates knock-on effects and eddying in otherwise insulated and isolated societies.

Second, lightning projects from the Global North, in this case the 40K Foundation, inject new perspectives into local communities from materials intermediaries. The boutique jewellery start-up, Roka, shows that there are opportunities to be had even in resource-constrained and hardship-rich places, such as in Bengaluru's stone quarries. The scope for grassroots innovation from external technical intermediaries (engineering undergraduates) is a case in point here, where an apparently useless material – stone dust – can be converted through expertise into marketable products that play off of the situated realities of the community, promoting the consumable values of hand-crafted artisanship and female empowerment.

Third, and in the same fashion as Chaphekar's line of 3D printers for the Indian market, products sensitive to the conditions of the Global South provide useful inputs into development projects. Jayesh Salvi's Mumbai-made design tool 3DTin solves many of the bottleneck issues for 3D4D: software licenses, technical training, incompatible file formats and inaccessible computing power, interfacing and storage.

Next in our exposition of the likely indicators for transition, we shall turn briefly to two current efforts to make 3D4D happen. Two forerunner projects – that is, iLab//Haiti and KIDmob and Dreambox

DOI: 10.1057/9781137365668.0010

Emergence and the Ethical Filament Foundation – bring together the different elements of the 3DP ecosystem into a coherent whole. These are forerunners of a functional 3D4D system.

Rabodé: making it work

In 2010, Haiti suffered a catastrophic earthquake that killed an estimated 250,000 people and crippled its already ailing infrastructure. In terms of large natural disasters, Haiti's earthquake tops the scale: it is 'the most destructive event a country has ever experienced when measured in terms of the number of people killed as a share of the country's population' and the assessment of damage to infrastructure is a staggering US$8.1 billion.[10]

Yet there are grassroots innovators working against these odds. Kate Ganim, Co-Director of KIDmob (a mobile, kid-centric design firm), is keen to stress that the 3DP initiative at iLab//Haiti is about much more than just the technology. Of course, Willio, Johnson, and Franky, the three young Haitians that form the 'maker' team, are being taught how to use, troubleshoot, and maintain the printers and how to use Makerware, AutoCAD, and 3D modelling tools like Rhino. However, the heart of the iLab//Haiti project is the core motivation to build an educational programme that, while providing practical skills for using the tools, equally focuses on learning how to think about the tools. According to Kate, the design process is an incredible tool for critical thinking and creative problem solving and many Haitians – whether culturally or out of necessity – are great at creative problem solving. This idea is captured by the Haitian term 'rabodé' which means 'make it work.' Rabodé is about resourcefulness and using what you have to get done what you need to get the job done. The iLab//Haiti's vision for the project is to apply rabodé beyond simple ad hoc band-aids. Instead, this team are thinking through rabodé as a solution for community action to formidable infrastructure deficits.

In addition to technical and design skills, the team are also being trained on business, business strategy and tools, and strategies for generally getting things done in resource-constrained conditions. As facilitators, they have a strong interest in teaching other Haitians for the long term benefits of Haiti's recovery. They are excited about teaching others how to use 3DP technology. By investing in people, Kate and her team

DOI: 10.1057/9781137365668.0010

believe that they can also serve an educational role that benefits generations of Haitians to come.

Of course, there are practical, infrastructural challenges to 3DP in post-disaster Haiti. The machines are housed at Haiti Communitere, where a generator produces electricity that is fairly consistent.[11] Power outages and really intense power surges are common in Haiti and so the project has installed two uninterruptible power supplies to help protect the equipment from power surges due to the infrastructural crisis. Whilst Internet connectivity is surprisingly reliable in the circumstances, albeit slow at times, the team have been learning 3D modelling skills so that they have the ability to generate models themselves, and will not have to rely solely on what is available online.

When asked about the popularity of the 3D Printing initiative in Haiti, Kate notes that there is a novelty and curiosity aspect to it: her impression is that the interest runs a lot deeper than surface experimentation. Some have expressed to iLab//Haiti that their work is a symbol of hope and pride. The idea here is that this new technology can be a bridge between Haiti and the rest of the world through communities of users that stretch beyond borders. The local media outlet, *Haiti Libre*, has realized that 3DP offers Haitians more than novelties:

> With its first two 3D printers 'Makerbots', iLab Haiti teaches how to model 3D objects, repair and maintain these machines... first productions are simple objects, disposable as clamps for umbilical cords often lacking in hospitals... Imagine tomorrow in Haiti, it will be possible for example to obtain to a mechanic a spare part manufactured on demand, without having to undergo the time of importation. Perspectives and the use of 3D printing are limited only by our imagination. A new revolution is underway in our country.[12]

On the subject of expansion, Kate is clear that this is in the hands of the Haitian team so they can establish the business that they want to run. Transitioning from a role of mentoring and support the team want to continue to help Haitians build the skills they need for everyday life. Kate asserts that with 3DP, iLab//Haiti have powerful tools and, thus, a powerful way of thinking, which are both of global value. Moreover, it is the community who stand to develop the operating skills and thus the innovation potential.

Kate is excited about the future possibilities for Haiti though. She sees the potential for Haitians to connect directly to a global economy. In this case, design and 3D modelling become a lot more important than 3D printing (other than perhaps test printing their 3D models). In this scenario, where local economies are incredibly weak, tapping into a strong global economy could have a lot more potential to generate revenue than selling locally. Conversely, there is the potential for design challenges in Haiti to be crowd sourced to the rest of the world – solutions could be implemented using the 3D printers.

From landfill to filament

At the beginning of this book, we presented a case study describing the 3D4D Challenge, which brought together innovators from all around the world – a first step in an intervention to assist in building a resilient sustainable community according to standards of comfort, convenience, and cleanliness in the Global North. So far we have surveyed 3D4D 'indicators' for separate elements within the 3DP ecosystem. Lastly, there are forerunners of what this ecosystem might be like as a system. In 2012, a glimmer of this scenario came into being with the manufacturing social enterprise Dreambox Emergence and the formation of the Ethical Filament Foundation. These are significant forerunners of 3D4D.

The hectic streets of Pune, India, bustle with energy, as motorbikes and auto-rickshaws careen below the sedate banyan trees whose hanging roots flaunt their freedom over the traffic congestion below. The colloquial Oxford of the East has confronted much change over the past half-century as population growth has accompanied the rise of India's middle-class and their offspring who move from nearby Mumbai and sometimes further afield to attend the spate of prestigious and not so-prestigious colleges that pepper the small city. What were once tree groves, large stately bungalows, and common pastures are now precariously high-rise residential blocks. Some of the college students never leave and stay local to create start-ups or work in the growing technology parks that provide gated and manicured havens for the talented.

It is in these conurbations where local companies jostle for the best real estate alongside India's up-and-coming ICT powerhouses: Satyam, Infosys and the older business houses known to all. These innovation hubs also provide sanctuary for more formal experiments in technology and engineering, such as the Ethical Filament Foundation, the brainchild of 3D4D Challenge finalists Just 3D Printing and Challenge organizers

DOI: 10.1057/9781137365668.0010

techfortrade. The hospitable and well-connected Pai family are trying something no one else has done before that goes utterly against the grain of their digital, network-heavy, and multinational residencies: a circular economy that connects communities at the base of India's pyramid with domestic and overseas early adopters of 3DP.

The story Jayant and Suchismita Pai tell in the tour of their home workshop of the founding of their idea betrays years of engagement at what appear to be different scales: first, volunteer work with local waste-picker communities at the coalface of India's rapid growth; and second, educational experience in the world's elite universities. Setting up a test 3DP workshop in their home garage – that icon of entrepreneurship alive and well in India too – the couple and their MIT-student son Sidhant along with Jayant's two expert engineering friends began to tinker with a 3DP cradle-to-grave concept inspired by similar innovators elsewhere. Just 3D Printing sources recycled plastics collected by local waste-pickers in Pune whose work operates at the very bottom of the pyramid on India's streets where every last resource, including all manner of waste from dung to plastic, represents an income. Specially designed machines will be used to transform this material into a filament suitable for use in 3D printers that can either be sold to Just 3D Printing, or to other users of this technology. For Just 3D Printing, the Goldilocks Zone will be found through the installation of low-cost DIY 3D printers into kiosks at popular locations across India that will be subsidized to give young entrepreneurs and students access to low-cost rapid prototyping. Local employees will receive relevant training sympathetic to their workloads and social status in order to staff the kiosks. These sites will not only be universities, print shops, cafes or corporations in the urban hubs, but schools, village community halls and, perhaps most importantly for 3D4D, waste disposal sites.

Suchismita had already gained the trust of a local union of waste-pickers, SWaCH, located at waste-sites across the city, which had lobbied for changes to regulations and facilities to provide some manner of protection to the workers at the lowest rung of the waste-picking hierarchy: women and children. This work with SWaCH is low-cost and scalable, empowering waste-pickers with the technology to produce extrusion thermo-plastic filament from high-density polyethylene (HDPE) waste. Visiting the site on a whirlwind tour it is no challenge to perceive the value of the improved conditions, educational programmes, innovative

DOI: 10.1057/9781137365668.0010

equipment, and other benefits of the union. Key contacts who work on these sites are also champions of this social innovation being taught how to use the custom 'Flakerbot' – built by Jayant and Sidhant – that shreds certain grades of plastic bottles into premium quality, international standard 3DP filament wire. Part of the genius of the cradle-to-grave innovation is its simplicity of use, with coloured operating buttons and ruggedized frame, perfect for the hard conditions of waste sites.

As they explain over lunch at one of Pune's many fashionable cafes, this trial will confirm the viability of a much more ambitious project: The Ethical Filament Foundation. The initiative will work in partnership with organizations around the world to set benchmarks for the manufacture of sustainable 3DP feedstocks made from recycled plastic waste in the Global South. The progress will be a recognized standard of wire that will be licensed for use by partner organizations and ultimately larger markets of sustainably minded 3DP users. Feedstock production from waste will be in accordance with The Ethical Filament Foundation's main goals. The Ethical Filament Foundation mark of quality will act as an assurance for companies and individual consumers wishing to purchase recycled filament.

techfortrade established the Ethical Filament Foundation in partnership with Just 3D Printing (now ProtoPrint) and Dreambox Emergence, which provides 3D printers for community based manufacturing in Guatemala as well as in the US at Michigan Technology University. The idea of Dreambox Emergence is to set up desktop-sized manufacturing stations in community hubs where local suppliers already operate. Dreambox produces objects (disability aids, water filters, and solar-power lamps) in response to demand for short runs of these specific products, thus putting into practice the ideals of 3D4D outlined in this book: no long-distance freight costs and no unwanted inventories of stock.

3DP dreams

In December 2013, techfortrade partnered with Dreambox Emergence to stage a 3D4D Challenge on the design site GrabCAD. The aim of the challenge was to crowdsource a design for children's sunglasses, to protect children living in Nebaj, Guatemala from Pterygium or Surfer's Eye. The Challenge generated an amazing array of entries and the designers' approaches ranged from using melted filament to attach the arms in place, to snap joints and clip on interchangeable designs. The judges were particularly interested in designs which would make the best use of 3D printing technology, ensuring ease of printing, structural security, and ease of assembly so that the frames were as close to ready to wear as possible. Designs that required the least amount of after-production assembly ensure that a crucial advantage of 3DP (a finished product virtually straight from the machine) can be utilized. The Challenge has proven to be a successful, initial step in the evolving process of using 3DP technology to its full potential in emerging markets.

3DP offers virtually limitless potential to produce the objects we view as every day necessities but which frequently prove prohibitively expensive or difficult to obtain in developing countries. This potential has often been dismissed as hype by the media and the opportunities posed for developing economies have been hampered by the price of the technology, but the 3D4D 2013 Challenge has proven that this technology can be used to address a specific problem and is moving on its way to providing tangible answers to economic supply problems in Guatemala and beyond. Most impressively, the Challenge has revealed a diverse and passionate community of designers and creative thinkers who are open to and can provide innovative solutions for problems around the globe, while 3DP provides the technology which links these designers and the people who can use their ideas together, with a result that can make a real difference in people's lives.

In December 2013, Dreambox Emergence set up their first centre in Guatemala and are now putting the finishing touches to their own 3D printer, specially developed for use in emerging markets where the machines are required to be robust and adaptable in harsher environments. Reliable electricity supply is a crucial obstacle in any developing country where technology is in use and the printers have been supplied with auxiliary supplies through battery packs to compensate for any interruption in delivery. In the future, the team is looking to incorporate solar panels and other forms of self-sustaining energy to make the printers as durable as possible.

Dreambox are aiming to start printing essential health, educational, and water supplies in March, (the finalist's designs being among those that will be available, courtesy of the integrated library of designs which is pre-installed in each printer). These designs include rulers for

DOI: 10.1057/9781137365668.0010

classrooms, tools such as valves and wrenches, water filters for sanitation and medical equipment, particularly eye wear.

Dreambox Emergence are at a very exciting point where they have adapted their technology to peak performance and can consider implementation across Guatemala and hopefully in the future, expanding out through Central America, Asia and Africa. 3DP offers the chance to print country and culturally specific items – providing resources and tools that apply in the countries where the machines will be based so that the technology is truly tailored and developed to address the specific needs of emerging markets and developing communities.

Providing the technology and the designs to create is only half the work needed to ensure that 3D printing is an asset in Nebaj and the wider developing world. Ensuring that communities understand how the machines work and why they are important is an area that the team in Nebaj are working hard to facilitate. They are currently training several members of the community in the maintenance of their 3D printers and use of the design software which will produce the objects. They are also working on an emerging market based curriculum which will educate the engineers who maintain the machines at present, and those who will learn how to use the devices in the future, ensuring that the knowledge and expertise are available for people to understand. Convincing the community that they can gain real benefits from mastering and using this technology has taken time and persistence, and the team are particularly looking forward to launching in March to see how the local community will take to the next stage of developing 3D printing in Nebaj. 3D printing offers the creation and production of essentials, but also the potential to design anything that might come to mind and ensuring that as many people as possible have access to the possibilities on offer.

This 3D4D Challenge 2013 has been a first step in what the team hopes will become a regular collaboration between the international design community and technology in use in developing countries. The initial puzzle of 3D printed sunglass frames has established that the community is present and willing, and that there is so much potential that it is now vital to build on the enthusiasm and creativity they have found from this challenge. Dreambox Emergence and techfortrade are always looking for the next expansion possible with 3D printing, especially given the enthusiasm they have accessed through such a vibrant and thoughtful design community.

DOI: 10.1057/9781137365668.0010

Notes

1 Gereffi, Wadhwa, Rissing and Ong (2008) 'Getting the Numbers Right: International Engineering Education in the United States, China, and India'.
2 Irani (2012) 'Print in 3D'.
3 Birtchnell (2013b) *Indovation: Innovation and a Global Knowledge Economy in India.*
4 Sethi (2014) 'Making 3D Printers Now Child's Play'.
5 Siyanbola, Egbetokun, Adebowale and Olamade (2012) 'Innovation Systems and Capabilities in Developing Regions: Concepts, Issues and Cases'.
6 Dumont (1972) *Homo Hierarchicus: The Caste System and its Implications.*
7 UNESCO (2013) 'Creative Economy Report 2013 Special Edition Widening Local Development Pathways': 89.
8 Harouni (2014) 'Lisa Harouni Thinks 3D Printing Can Make Design Accessible to You'.
9 Kelly (2008) 'Becoming Screen Literate'.
10 Cavallo, Powell and Becerra (2010) 'Estimating the Direct Economic Damages of the Earthquake in Haiti'.
11 http://haiti.communitere.org.
12 Haiti Libre (2014) 'Haiti – Technology: 3D Printing Makes Its Entry into Haiti'.

DOI: 10.1057/9781137365668.0010

Conclusion

Abstract: *The chapters in this book offer a mandatory review of one of the most significant technological innovations of the early twenty-first century: 3D printing. Birtchnell and Hoyle appraise the central arguments of the book in this final chapter and revisit the development implications of 3D printing in light of the case studies presented in Chapters 5–6. In order to bolster sustainable communities in the Global South, 3D printing would need mediation and careful refinement according to third party standards so that it would not exacerbate poverty further. Although upbeat about 3D printing's potential for development action, Birtchnell and Hoyle are also critical about how this social transformation might take place and emphasize that there need to be ground rules in the ordering of the various elements in the 3D printing ecosystem so that it is 'just right' for development.*

Birtchnell, Thomas and William Hoyle. *3D Printing for Development in the Global South: The 3D4D Challenge.* Basingstoke: Palgrave Macmillan, 2014.
DOI: 10.1057/9781137365668.0011.

In this short, pamphlet-scale book on the development potential of 3DP, we have been at pains to avoid prescribing this new process of producing objects as some sort of universal solution to the world's problems. We have tried to show that 3DP is on a par with other technologies that are for all intents and purposes reaching ubiquity in the Global South: the mobile phone and automobile.

Instances of technological innovation are rarely, if ever, a 'magic bullet' for large-scale social transformation. Instead, examples from the developing world's adoption of network technologies, including the Internet and mobile phone, indicate that there is potential for various niches to cascade into wider social landscapes and enable the inclusivity challenge to be tackled by individuals and their own creativity. The innovators pooled together in the 3D4D Challenge demonstrate that the range of applications that 3DP can contribute to is wide and touches on those areas identified by the UN as important development goals.

Whether it is farmers seeking crop data, itinerant labourers pursuing informal jobs, or would-be entrepreneurs needing micro-finance loans, technological innovation has been able to empower their interests, often on a case-by-case basis. To us, 3DP appears very much like these other technological innovations in its potential for mass appeal in the Global South, its industry support, and its burgeoning open source movement, with the RepRap being the most obvious exemplar. The 3DP indicators described in Chapters 5–6 of this book fortify this assessment. Moreover, there are forerunners, also acknowledged in this book operating now in the Global South that offer an insight into how social change that raises living standards and is driven by 3DP could take place.

As we remarked in Chapter 2, 3DP is fertile territory for reconceptualizing the production and consumption regimes that provide people with the objects they use in their everyday lives. This epochal shift would be felt most in how people make their livings and establish norms of comfort, convenience, and cleanliness that compare with standards set in the Global North. We reviewed the various provocations and commentaries on a third industrial revolution and reflected on the nature of the current manufacturing system and its maturation through new shades of production (post-Fordism) and consumerism (prosumerism) in Chapter 3.

Furthermore, a 'Goldilocks Zone' that is 'just right' for development was pegged to various standards that point to a hot spot for 3DP's mass

DOI: 10.1057/9781137365668.0011

adoption. Many of the criticisms of 3DP were also broached and set aside as 'first world problems' not necessarily appropriate in the Global South.

Academic literature describes the growing promise of bottom-up efforts that are self-sufficient from the top-down 'big development' of mainstream governmental and intergovernmental interventions. In Chapter 4, we gave thought to how 3DP has already changed the development game in its first flush as a niche innovation. A 3D4D wishlist was proposed that would allow a grassroots innovation to find scale and become consequential.

In Chapters 5–6, we illustrated our careful appraisal of 3DP and its global consequences with case studies of various pragmatic examples of the deployment of this innovation at a grassroots level. These mostly first-hand accounts of 3D4D target specific communities in the Global South and attempt to balance out some of the extant inequalities symptomatic of the dominant worldwide production and consumption system. Enticingly, these examples synchronize with much wider social movements that also destabilize this system reliant on global production networks (GPNs) and internationally variable dimensions of austerity. 3D4D was imagined as being optimized through community printers, open repositories, recycled materials, and sympathetic infrastructure.

Further support is needed for initiatives that inject vigour into community action across territorial and political divides. The 3D4D Challenge, showcased here in this book, was one such effort to mobilize resources and nurture debate about how the various elements (niches in technologies, designs, materials, and infrastructures) in what we called the 3DP ecosystem, will come together to occupy a configuration that is just right for sustainable development. We acknowledge that this Goldilocks Zone is still not entirely there yet and work remains to bring partisan elements into a fruitful alignment. Nevertheless, we are critically upbeat about the demonstrated assiduity of the various innovators in the 3D4D Challenge to meaningful community action through the rubric of 3D4D. Of most promise is the idea of circular economies that allow accountability and transparency to be injected into a production and consumption system that is at the moment notoriously opaque. If 3D4D has made some headway in securing resolve for greater equality and accountability in consumption and production, then the efforts of the many diverse agents operating across the Global South who featured in this book will prove gainful for development.

DOI: 10.1057/9781137365668.0011

Bibliography

Abrol, D. (2004) 'Lessons from the Design of Innovation Systems for Rural Industrial Clusters in India', *Asian Journal of Technology Innovation*, 12 (2) pp. 67–97.

Adama, O. (2012) 'Urban Livelihoods and Social Networks: Emerging Relations in Informal Recycling in Kaduna, Nigeria', *Urban Forum*, 23 (4) pp. 449–66.

Agar, J. (2013) *Constant Touch: A Global History of the Mobile Phone* (London: Icon Books, Limited).

Anand, P. B. (2000) 'Co – Operation and the Urban Environment: An Exploration', *The Journal of Development Studies*, 36 (5) pp. 30–58.

Anderson, C. (2012) *Makers: The New Industrial Revolution* (London: Random House).

Anzalone, G. C., Zhang, C., Wijnen, B., Sanders, P. G. & Pearce, J. M. (2013) 'A Low-Cost Open-Source Metal 3-D Printer', *IEEE Access*, 5 December.

Araujo, L. (2013) 'What Have Markets Ever Done for the Poor?', *Marketing Theory*, 13 (3) pp. 385–88.

Atal, Y. (1972) 'Role of Values and Institutions', in Fonseca, A. J. ed. *Challenge of Poverty in India* (Delhi: Vikas) pp. 87–104.

Bapuji, H. (2012) *Not Just China: The Rise of Recalls in the Age of Global Business* (Basingstoke: Palgrave Macmillan).

Beausang, F. (2012) *Globalization and the Brics: Why the Brics Will Not Rule the World for Long* (Basingstoke: Palgrave Macmillan).

Beer, D. & Burrows, R. (2013) 'Popular Culture, Digital Archives and the New Social Life of Data', *Theory Culture Society*, pp. 1–25.

DOI: 10.1057/9781137365668.0012

Benvenuti, S., Ceccanti, F. & De Kestelier, X. (2013) 'Living on the Moon: Topological Optimization of a 3D-Printed Lunar Shelter', *Nexus Network Journal*, 15 (2) pp. 285–302.

Bergmann, L. (2013) 'Bound by Chains of Carbon: Ecological – Economic Geographies of Globalization', *Annals of the Association of American Geographers*, 103 (6) pp. 1348–70.

Berman, B. (2012) '3-D Printing: The New Industrial Revolution', *Business Horizons*, 55 (2) pp. 155–62.

Birtchnell, T. (2012) 'Elites, Elements and Events: Practice Theory and Scale', *Journal of Transport Geography*, 24 pp. 497–502.

Birtchnell, T. (2013a) 'Fill the Ships and We Shall Fill the Shops: The Making of Geographies of Manufacturing', *Area*, 45 (4) pp. 436–42.

Birtchnell, T. (2013b) *Indovation: Innovation and a Global Knowledge Economy in India* (Basingstoke: Palgrave Macmillan).

Birtchnell, T. (2013c) *Indovation: Innovation in India's Knowledge Economy* (Basingstoke: Palgrave Macmillan).

Birtchnell, T. & Urry, J. (2013) '3D, SF and the Future', *Futures*, 50 pp. 25–34.

Birtchnell, T. & Urry, J. (2014) 'The Mobilities and Post-Mobilities of Cargo', *Consumption Markets & Culture*, Forthcoming.

Birtchnell, T., Viry, G. & Urry, J. (2013) 'Elite Formation in the Third Industrial Revolution', in Birtchnell, T. & Caletrío, J. eds *Elite Mobilities* (Abingdon: Routledge) pp. 62–77.

Brainard, L. (2009) *Global Development 2.0: Can Philanthropists, the Public, and the Poor Make Poverty History?* (New York: Brookings Institution Press).

Bruns, A. & Schmidt, J-H (2011) 'Produsage: A Closer Look at Continuing Developments', *New Review of Hypermedia and Multimedia*, 17 (1) pp. 3–7.

Cano, J. L. C. (2011) 'The Cambrian Explosion of Popular 3D Printing', *International Journal of Artificial Intelligence and Interactive Multimedia*, 1 (4) pp. 31–33.

Castree, N., Coe, N. M., Ward, K. & Samers, M. (2004) *Spaces of Work: Global Capitalism and Geographies of Labour* (London: SAGE Publications).

Cavallo, E., Powell, A. & Becerra, O. (2010) 'Estimating the Direct Economic Damages of the Earthquake in Haiti', *The Economic Journal*, 120 (546) pp. F298–F312.

Chronopoulos, T. (2006) 'The Cartoneros of Buenos Aires, 2001–2005 1', *City* 10 (2) pp. 167–182.

DOI: 10.1057/9781137365668.0012

Clark, N., Frost, A., Maudlin, I. & Ward, A. (2013) *Technology Development Assistance for Agriculture Putting Research into Use in Low Income Countries* (Abingdon: Routledge).

Coe, N. M. & Hess, M. (2013) 'Global Production Networks, Labour and Development', *Geoforum*, 44 pp. 4–9.

Coe, N. M., Hess, M., Yeung, H. W.-c, Dicken, P. & Henderson, J. (2004) '"Globalizing" Regional Development: A Global Production Networks Perspective', *Trans Inst Br Geogr*, 29 pp. 468–84.

Comor, E. (2010) 'Contextualizing and Critiquing the Fantastic Prosumer: Power, Alienation and Hegemony', *Critical Sociology*, 37 (3) pp. 309–27.

Cresswell, T. & Martin, C. (2012) 'On Turbulence: Entanglements of Disorder and Order on a Devon Beach', *Tijdschrift voor economische en sociale geografie*, 103 (5) pp. 516–29.

D'Aveni, R. A. (2013) '3-D Printing Will Change the World', *Harvard Business Review* (Harvard: Harvard Business Publishing) March. http://www.hbr.org/2013/03/3-d-printing-will-change-the-world/.

Dauvergne, P. & LeBaron, G. (2012) 'The Social Cost of Environmental Solutions', *New Political Economy*, 18 (3) pp. 410–30.

Day, P. (2011) 'Will 3D Printing Revolutionise Manufacturing?' (http://www.bbc.co.uk/news/business-14282091) Accessed 3 January.

Dean, W. (2012) 'What's the Big Deal with 3D Printing?', *The Independent* (London: Independent Print Ltd) 25 October 2012. http://www.independent.co.uk/life-style/gadgets-and-tech/features/whats-the-big-deal-with-3d-printing-8225267.html.

Donner, J. (2007) 'The Rules of Beeping: Exchanging Messages Via Intentional "Missed Calls" on Mobile Phones', *Journal of Computer-Mediated Communication*, 13 (1) pp. 1–22.

Doyle, R., Froede, E., Saint John, D. & Devon, R. (2010) 'In the Beginning Was the NoöSphere: Community and Collaboration in Open Source Evolution of Technology' (http://www.asee.org/documents/sections/middle-atlantic/fall-2010/01-Understanding-Open-Source-Design-A-White-Paper.pdf) ASEE Proceedings.

Dumont, L. (1972) *Homo Hierarchicus: The Caste System and Its Implications* (London: Paladin).

Durkheim, E. (1997) *The Division of Labor in Society* (London: Free Press).

Dvorak, P. (2006) 'Rapid Prototypes for Baghdad, One Year Later' (http://medicaldesign.com/prototyping/rapid-prototypes-baghdad-one-year-later) Accessed 17 August.

DOI: 10.1057/9781137365668.0012

Easton, T. A. (2011) 'A Recession in the Economy Trust', in Ricci, G. ed. *Values and Technology: Religion and Public Life* (New Jersey: Transaction Publishers) pp. 159–69.

Econolyst (2012) 'The 3D4D Challenge Explained' (http://youtu.be/DYorQdFC4TU) Accessed 2 February.

European Commission (2012) 'Industrial Revolution Brings Industry Back to Europe' (Brussels: European Commission, Memo/12/759).

Foresight (2010) 'Technology and Innovation Futures: UK Growth Opp.ortunities for the 2020s' (London: The Goverment Office for Science). http://www.bis.gov.uk/assets/foresight/docs/general-publications/10-1252-technology-and-innovation-futures.pdf.

Foresight (2013) 'The Future of Manufacturing: A New Era of Opp. ortunity' (London: The Government Office for Science). https://http://www.gov.uk/government/uploads/system/uploads/attachment_data/file/255922/13-809-future-manufacturing-project-report.pdf.

Fox, S. (2013) 'Open Prosperity: Breaking Down Financial and Educational Barriers to Creating Physical Goods', *IEEE Technology and Society Magazine*, 9 December. http://ieeexplore.ieee.org/stamp/stamp.jsp?arnumber=06679306.

Fréchette, L. (2010) 'Foreword', in Clapp., J. & Wilkinson, R. eds *Global Governance, Poverty and Inequality* (Abingdon: Routledge) pp. xxi–xxviii.

Gada, K. (2011) 'Reducing Poverty through Personal Manufacturing' (http://www.ssireview.org/blog/entry/reducing_poverty_through_personal_manufacturing) Accessed 20 August.

George, S. (1986) *How the Other Half Dies: The Real Reasons for World Hunger* (Harmondsworth: Penguin).

Gereffi, G., Wadhwa, V., Rissing, B. & Ong, R. (2008) 'Getting the Numbers Right: International Engineering Education in the United States, China, and India', *Journal of Engineering Education*, 97 (1) pp. 13–25.

Gershenfeld, N. (2007) *Fab: The Coming Revolution on Your Desktop – from Personal Computers to Personal Fabrication* (New York: Basic Books).

Ghosh, R. A. (2003) 'Licence Fees and GDP Per Capita: The Case for Open Source in Developing Countries', *First Monday*, 8 (12).

Gibson, C., Farbotko, C., Gill, N., Head, L. & Waitt, G. (2013) *Household Sustainability: Challenges and Dilemmas in Everyday Life* (Cheltenham: Edward Elgar).

Giddens, A. (2012) 'Re-Industrialization in a New Electric Age' (http://youtu.be/O-CQcOxQZDI) Accessed 12 December.

Giddens, A. (2013) 'Lord Giddens on 3D Printing' (http://youtu.be/m4Kz6Mdqypc) Accessed 15 December.

Gordon, L. (2011) 'Rapid Prototyping for the Masses', *Machine Designs*, 83 (10) pp. 40–42.

Guess, C. (2013) '3D Printing for Africa's Development' (http://www.seadiaspora.com/2013/03/author-christopher-guess-author.html - .Ux_0815RF1R) Accessed 3 March.

Gupta, D. (2010) *The Caged Phoenix: Can India Fly?* (Chicago: Stanford University Press).

Haiti Libre (2014) 'Haiti – Technology: 3D Printing Makes Its Entry into Haiti', *Haiti Libre* (Port-au-Prince: Haitilibre.com). http://www.haitilibre.com/en/news-10243-haiti-technology-3d-printing-makes-its-entry-into-haiti.html.

Hargreaves, T., Hielscher, S., Seyfang, G. & Smith, A. (2013) 'Grassroots Innovations in Community Energy: The Role of Intermediaries in Niche Development', *Global Environmental Change*, 23 (5) pp. 868–80.

Harouni, L. (2014) 'Lisa Harouni Thinks 3D Printing Can Make Design Accessible to You' (http://youtu.be/opoTmFlpcoE) Accessed 8 October.

Haseloff, A. M. (2005) 'Cybercafés and Their Potential as Community Development Tools in India', *The Journal of Community Informatics*, 1 (3).

Hayami, Y., Dikshit, A. K. & Mishra, S. N. (2006) 'Waste Pickers and Collectors in Delhi: Poverty and Environment in an Urban Informal Sector', *The Journal of Development Studies*, 42 (1) pp. 41–69.

Haynes, D. E. (2013) 'The Making of the Hyper-Industrial City in Western India: The Transformation of Artisanal Towns into Middle-Sized Urban Centres, 1930–1970', *South Asia: Journal of South Asian Studies*, 36 (3) pp. 336–53.

Heeks, Richard (2009) 'The ICT4D 2.0 Manifesto:Where Next for ICTs and International Development?' (Manchester: Institute for Development Policy and Management, University of Manchester). http://www.sed.manchester.ac.uk/idpm/research/publications/wp/di/index.htm

Heyer, J. (2013) 'Integration into a Global Production Network: Impacts on Labour in Tirupp.ur's Rural Hinterlands', *Oxford Development Studies*, 41 (3) pp. 307–21.

Hopkinson, N., Hague, R. J. M. & Dickens, P. M. (2006) 'Introduction to Rapid Manufacturing', in Hopkinson, N. & Hague, R. J. M. eds *Rapid Manufacturing: An Industrial Revolution for the Digital Age* (Chichester: John Wiley & Sons) pp. 1–4.

Hu, S. J. (2013) 'Evolving Paradigms of Manufacturing: From Mass Production to Mass Customization and Personalization', *Procedia CIRP*, 7 pp. 3–8.

Hulme, D. (2010) 'Governing Global Poverty? Global Ambivalence and the Millennium Development Goals', in Clapp., J. & Wilkinson, R. eds *Global Governance, Poverty and Inequality* (Abingdon: Routledge) pp. 135–61.

Hunt, C. (2013) 'Formalization Beckons: A Baseline of Informal Recycling Work in Buenos Aires', *Environment and Urbanization*, 25 pp. 501–21.

Ilan, Z. B. (2011) 'From Economy of Commodities to Economy of Ideas: Hardware as Social Medium', *Design Management Review*, 22 (3) pp. 44–53.

Irani, M. (2012) 'Print in 3D', *Sunday Times of India* (Mumbai: 22 January). http://epaper.timesofindia.com/Default/Layout/Includes/TOINEW/ArtWin.asp?App.Name=1&BaseHref=TOIM/2012/01/22&ViewMode=HTML&EntityId=Ar02300&From=Archive&PageLabel=23&Skin=TOINEW&Source=Page.

Jeffrey, C. (2010) *Timepass: Youth, Class, and the Politics of Waiting in India* (Stanford: Stanford University Press).

Jeffrey, C. & Dyson, J. (2013) 'Zigzag Capitalism: Youth Entrepreneurship in the Contemporary Global South', *Geoforum*, 49 pp. R1–R3.

Jeffrey, C. & Young, S. (2013) 'JugāD: Youth and Enterprise in India', *Annals of the Association of American Geographers*, 104 (1) pp. 182–95.

Jessop, B. & Sum, N-L (2006) *Beyond the Regulation App.roach: Putting Capitalist Economies in Their Place* (Cheltenham: Edward Elgar).

Jobson, C. (2011) 'Markus Kayser Builds a Solar-Powered 3D Printer That Prints Glass from Sand and a Sun-Powered Cutter' (http://www.thisiscolossal.com/2011/06/markus-kayser-builds-a-solar-powered-3d-printer-that-prints-glass-from-sand-and-a-sun-powered-laser-cutter/) Accessed 3 September.

Kadushin, R. (2013) 'Bearina Iud Concept' (http://www.ronen-kadushin.com/index.php/open-design/bearina-iud-concept/) Accessed 6 Feburary.

DOI: 10.1057/9781137365668.0012

Karnani, A. (2007) 'Doing Well by Doing Good—Case Study: "Fair & Lovely" Whitening Cream', *Strategic Management Journal*, 28 pp. 1351–57.

Kaufmann, V. (2013) 'The Dharavi Paradox' (http://urbz.net/the-dharavi-paradox/) Accessed 3 Feburary.

Kelly, K. (2008) 'Becoming Screen Literate', *The New York Times Magazine* (New York: The New York Times Company) 21 November. http://www.nytimes.com/2008/11/23/magazine/23wwln-future-t.html?pagewanted=all&_r=0.

Kemp, R. & van Lente, H. (2013) 'The Dual Challenge of Sustainability Transitions: Different Trajectories and Criteria', in Cohen, M. J. ed. *Innovations in Sustainable Consumption: New Economics, Socio-Technical Transitions and Social Practices* (Cheltenham: Edward Elgar) pp. 115–32.

Koebler, J. (2013) 'Is 3D Printing the Future of Disaster Relief?', *Motherboard* (New York: Vice Media Inc.). http://motherboard.vice.com/blog/is-3d-printing-the-future-of-disaster-relief.

Kostakis, V. (2013) 'At the Turning Point of the Current Techno-Economic Paradigm: Commons-Based Peer Production, Desktop Manufacturing and the Role of Civil Society in the Perezian Framework', *TripleC*, 11 (1) pp. 173–90.

Lash, S. & Urry, J. (1994) *Economies of Signs and Space* (London: SAGE Publications Limited).

Lipson, H. (2011) 'This Will Change Everything', *New Scientist* (New Scientist). http://www.newscientist.com/issue/2823.

Madon, S. (2009) *E-Governance for Development: A Focus on Rural India* (Basingstoke: Palgrave Macmillan).

Markillie, P. (2012) 'A Third Industrial Revolution' (http://www.economist.com/node/21552901) Accessed 14 September.

Mayhew, H. (1851) *London Labour and the London Poor* (London: Griffin, Bohn and Company).

McGann, J. G. & Sabatini, R. (2011) *Global Think Tanks: Policy Networks and Governance* (Abingdon: Routledge).

Mercer, C. (2006) 'Telecentres and Transformations: Modernizing Tanzania through the Internet', *African Affairs*, 105 (419) pp. 243–64.

Michael, M. (2013) 'Process and Plasticity: Printing, Prototyping and the Process of Plastic', in Gabrys, J., Hawkins, G. & Michael, M. eds *Accumulation: The Material Politics of Plastic* (Abingdon: Routledge) pp. 30–46.

DOI: 10.1057/9781137365668.0012

Mitra, S. (2010) *Vision India 2020* (BookSurge Publishing).
Mudliar, P. & Pal, J. (Year) 'ICTD in the Popular Press: Media Discourse around Aakash, the "World's Cheapest Tablet" ', *ICTD*, Cape Town, South Africa. Institution. Department. 7–10 December.
Mwesige, P. G. (2004) 'Cyber Elites: A Survey of Internet Café Users in Uganda', *Telematics and Informatics*, 21 (1) pp. 83–101.
Nasman, C. (2013) 'Dutch Architects Say 3D Print Technology Could Solve the World's Housing Problems' (http://www.dw.de/dutch-architects-say-3d-print-technology-could-solve-the-worlds-housing-problems/a-17036716) Accessed 3 February.
Nathan, M. (2011) 'Solar-Powered RepRap Prints Even When the Power Is Out' (http://hackaday.com/2011/09/15/solar-powered-reprap-prints-even-when-the-power-is-out/) Accessed 23 April.
Naughton, J. (2012) 'Is 3D Printing the Key to Utopia?', *The Observer* (Guardian News and Media Limited), 13 April 2012. http://www.guardian.co.uk/technology/2012/may/13/3d-printing-digital-manufacturing-industry.
Neagle, C. (2013) '3D Printing: Good for Retail? Tesco Cio Says So', *Network World* (Framingham: Network World). http://www.networkworld.com/news/2013/092713-3d-printing-retail-274293.html.
Newman, S. (1981) 'Newspapers on Computers' (San Francisco: KRON). http://youtu.be/5WCTn4FljUQ
Nnorom, I. C. & Osibanjo, O. (2008) 'Overview of Electronic Waste (E-Waste) Management Practices and Legislations, and Their Poor App.lications in the Developing Countries', *Resources, Conservation and Recycling*, 52 (6) pp. 843–58.
Nye, D. E. (2007) *Technology Matters: Questions to Live With* (Cambridge MA: MIT Press).
Nye, D. E. (2013) *America's Assembly Line* (Cambridge, Mass.: MIT Press).
Olinto, P. & Uematsu, H. (2010) 'The State of the Poor: Where Are the Poor and Where Are They Poorest?'. World Bank. Poverty Reduction and Equity Department.
Pearce, J. M. (2012) 'The Case for Open Source App.ropriate Technology', *Environment, Development and Sustainability*, 14 (3) pp. 425–31.
Polanyi, K. (2001) *The Great Transformation: The Political and Economic Origins of Our Time* (New York: Beacon Press).

DOI: 10.1057/9781137365668.0012

Potstada, M. & Zybura, J. 'The Role of Context in Science Fiction Prototyping: The Digital Industrial Revolution', *Technological Forecasting and Social Change*, (0).

Prahalad, C. K. (2010) *The Fortune at the Bottom of the Pyramid: Eradicating Poverty through Profits* (New York: Wharton School Publishing).

Ramos, J. M. (2013) 'The Future of Power in the Network Era', *Journal of Futures Studies*, 17 (4) pp. 71–92.

Rangaswamy, N. & Densmore, M. (2013) 'Understanding Jugaad: ICTD and the Tensions of App.ropriation, Innovation and Utility' *Proceedings of the Sixth International Conference on Information and Communications Technologies and Development: Notes - Volume 2*, Cape Town, South Africa Institution. Department. pp. 120–23.

Rao, S. S. (2003) 'Information Systems in Rural Indian Communities', *Journal of Computer Information Systems*, 44 (1) pp. 48–56.

Reeves, P. (2013a) '3D4D Challenge Finalist Rehearsal Day' (http://vimeo.com/52226305) Accessed 2014.

Reeves, P. (2013b) 'Jam Tomorrow: The Realities of 3D Printing Today' (http://www.econolyst.co.uk) Accessed 6 February.

Richardson, M., Elliott, S. & Haylock, B. (2013) 'This Home Is a Factory: Implications of the Maker Movement on Urban Environments', *Craft + Design Enquiry*, (5) pp. 1–3.

Rifkin, J. (2011) *The Third Industrial Revolution: How Lateral Power Is Transforming Energy, the Economy, and the World* (Basingstoke: Palgrave Macmillan).

Rigi, J. (2013) 'Peer Production and Marxian Communism: Contours of a New Emerging Mode of Production', *Capital & Class*, 37 (3) pp. 397–416.

Ritzer, G. & Jurgenson, N. (2010) 'Production, Consumption, Prosumption: The Nature of Capitalism in the Age of the Digital "Prosumer"', *Journal of Consumer Culture*, 10 (1) pp. 13–36.

Rivera, J. & van der Meulen, R. (2013) 'Gartner Says Worldwide Shipments of 3D Printers to Grow 49 Percent in 2013' (http://www.gartner.com/newsroom/id/2600115) Accessed 3 March.

Roberson, D. A., Espalin, D. & Wicker, R. B. (2013) '3D Printer Selection: A Decision-Making Evaluation and Ranking Model', *Virtual and Physical Prototyping*, 8 (3) pp. 201–12.

Rogers, A., Castree, N. & Kitchin, R. (2013) *A Dictionary of Human Geography* (Oxford: Oxford University Press).

DOI: 10.1057/9781137365668.0012

Satterthwaite, D. & Mitlin, D. (2014) *Reducing Urban Poverty in the Global South* (Abingdon: Routledge).

Schwittay, A. F. (2013) 'Making Poverty into a Financial Problem: From Global Poverty Lines to Kiva.Org', *Journal of International Development*, 26 (4) pp. 508–19.

Science Museum (2013) 'Chris Tuck on Why 3D Printing Is Not "Instant"' (http://youtu.be/MT1-ftYA5wM) Accessed 3 March.

Sethi, A. (2014) 'Making 3D Printers Now Child's Play' (http://timesofindia.indiatimes.com/india/Making-3D-printers-now-childs-play/articleshow/29743604.cms) Accessed 3 February.

Seyfang, G. & Longhurst, N. (2013) 'Desperately Seeking Niches: Grassroots Innovations and Niche Development in the Community Currency Field', *Global Environmental Change*, 23 (5) pp. 881–91.

Shapira, P., Youtie, J. & Urmanbetova, A. (2004) 'Advanced Technology and the Future of U.S. Manufacturing' (Atlanta: SRI International), School of Public Policy and the Georgia Tech Economic Development Institute Georgia Institute of Technology. https://smartech.gatech.edu/bitstream/handle/1853/41854/101112_gtworkshop-task1e-summary-report-M-26-6H2-1.pdf.

Shaw, D. (2012) 'Bras and Guitars: What People Today Are 3D Printing' (http://www.bbc.com/news/technology-20047744) Accessed 23 October.

Shove, E. (2003) *Comfort, Cleanliness and Convenience: The Social Organization of Normality* (Oxford: Berg).

Shove, E., Pantzar, M. & Watson, M. (2012) *The Dynamics of Social Practice: Everyday Life and How It Changes* (London: SAGE Publications Limited).

Singh, T. (2013) 'West African Inventor Makes a $100 3D Printer from E-Waste' (http://inhabitat.com/west-african-inventor-makes-a-100-3d-printer-from-e-waste/) Accessed 10 December.

Sissons, A. & Thompson, S. (2012) 'Three Dimensional Policy: Why Britain Needs a Policy Framework for 3D' (London: Big Innovation Centre), 16 October 2012. http://biginnovationcentre.com/Reports/23/Three-Dimensional-Policy-Why-Britain-needs-a-policy-framework-for-3D.

Siyanbola, W., Egbetokun, A. E., Adebowale, B. & Olamade, O. (2012) 'Innovation Systems and Capabilities in Developing Regions: Concepts, Issues and Cases', in Siyanbola, W., Egbetokun, A. E., Adebowale, B. A. & Olamade, O. eds *Innovation Systems and*

DOI: 10.1057/9781137365668.0012

Capabilities in Developing Regions: Concepts, Issues in Developing Regions (Farnham: Gower) pp. 1–12.

Slade, G. (2006) *Made to Break: Technology and Obsolescence in America* (Harvard: Harvard University Press).

Smiles, S. (1866) *Self Help* (London: John Murray).

Smith, A. (2011) 'Viewpoints: "What Do You Think Should Be the Two or Three Highest Priority Political Outcomes of the United Nations Conference on Sustainable Development (Rio+20), Scheduled for Rio De Janeiro in June 2012?"', *Natural Resources Forum*, 35 pp. 334–42.

Smith, A. & Seyfang, G. (2013) 'Constructing Grassroots Innovations for Sustainability', *Global Environmental Change*, 23 (5) pp. 827–29.

Sovacool, B. K., Dhakal, S., Gipp.ner, O. & Jain Bambawale, M. (2013) 'Peeling the Energy Pickle: Expert Perceptions on Overcoming Nepal's Electricity Crisis', *South Asia: Journal of South Asian Studies*, 36 (4) pp. 496–519.

Srivastava, I. (2013) 'Worldwide Shipments of 3D Printers to Grow 49% in 2013: Gartner', *The Times of India* (Chennai: Bennett, Coleman & Co. Ltd.), 3 October. http://timesofindia.indiatimes.com/business/india-business/Worldwide-shipments-of-3D-printers-to-grow-49-in-2013-Gartner/articleshow/23470424.cms.

Stahl, F. (2013) *Worker Leadership: America's Secret Weapon in the Battle for Industrial Competitiveness* (Cambridge, Mass.: MIT Press).

Sumner, A. & Mallett, R. (2012) *The Future of Foreign Aid: Development Co-Operation and the New Geography of Global Poverty* (Basingstoke: Palgrave Macmillan).

Taleb, N. N. (2008) *The Black Swan: The Impact of the Highly Improbable* (London: Penguin Books Limited).

Tang, T., Karhu, K. & Hämäläinen, M. (2011) 'Community Innovation in Sustainable Development: A Cross Case Study', *World Academy of Science, Engineering and Technology*, 49 pp. 396–403.

The Economist (2012a) '3D Printing: A Third-World Dimension', *The Economist* (Seattle: The Economist Newspaper Limited), 3 November 2012. http://www.economist.com/news/science-and-technology/21565577-new-manufacturing-technique-could-help-poor-countries-well-rich-ones.

The Economist (2012b) 'A Third-World Dimension' (http://www.economist.com/news/science-and-technology/21565577-new-manufacturing-technique-could-help-poor-countries-well-rich-ones) Accessed 3 November.

DOI: 10.1057/9781137365668.0012

Tsing, A. L. (2012) 'Empire's Salvage Heart: Why Diversity Matters in the Global Political Economy', *Focaal — Journal of Global and Historical Anthropology*, 64 pp. 36–50.

Tuck, C. & Hague, R. (2006) 'The Pivotal Role of Rapid Manufacturing in the Production of Cost-Effective Customised Products', *International Journal of Mass Customisation*, 1 (2) pp. 360–73.

UN (2012) 'UN System Task Team on the Post-2015 Development Agenda' (New York: Department of Economic and Social Affairs and the United Nations Development Programme). http://www.un.org/en/development/desa/policy/untaskteam_undf/thinkpieces/11_ips_science_innovation_technology.pdf) 18 March 2014.

UNEP (2012) 'Vital Waste Graphics 3' (Geneva: Secretariat of the Basel Convention prepared by Zoï Environment Network and GRID-Arendal).

UNESCO (2013) 'Creative Economy Report 2013 Special Edition Widening Local Development Pathways' (New York: (UNDP), UNDP) United Nations.

UNESCO (2014) 'Education for All Global Monitoring Report' (Paris: EFA Monitoring Report Team UNESCO). http://www.unesco.org/new/en/education/themes/leading-the-international-agenda/efareport/reports/.

Urry, J. (2007) *Mobilities* (Cambridge: Polity).

Urry, J. (2013) *Societies Beyond Oil: Oil Dregs and Social Futures* (London: Zed).

Urry, J. (2014) *Offshoring* (Cambridge: Polity).

Vannini, P. & Taggart, J. (2013) 'Domestic Lighting and the Off-Grid Quest for Visual Comfort', *Environment and Planning D: Society and Space*, 31 (6) pp. 1076–90.

Velazco, C. (2013) 'Motorola Taps 3D Systems to Produce Parts for Its Crazy Modular Smartphones', *TechCrunch* (New York: AOL Inc.), 22 November. http://techcrunch.com/2013/11/22/motorola-taps-3d-systems-to-produce-parts-for-its-crazy-modular-smartphones/.

Venta, L., Isomursu, M., Ahtinen, A. & Ramiah, S. (2008) '"My Phone Is a Part of My Soul" - How People Bond with Their Mobile Phones', *Mobile Ubiquitous Computing, Systems, Services and Technologies, 2008. UBICOMM '08. The Second International Conference on* Institution. Department. 29 September–4 October 2008, pp. 311–17.

Vinodh, S., Sundararaj, G., Devadasan, S. R., Kuttalingam, D. & Rajanayagam, D. (2009) 'Agility through Rapid Prototyping

DOI: 10.1057/9781137365668.0012

Technology in a Manufacturing Environment Using a 3D Printer', *Journal of Manufacturing Technology Management*, 20 (7) pp. 1023–41.

Wallop, H. (2013) 'With 3D Printers, You Can Change the World (or at Least Mend a Train Set)', *The Telegraph* (London: Telegraph Media Group), 4 October. http://www.telegraph.co.uk/technology/news/10356352/With-3D-printers-you-can-change-the-world-or-at-least-mend-a-train-set.html.

Whitwam, R. (2012) 'Why the 3D Printing Revolution Won't Happen in Your Garage' (http://www.extremetech.com/extreme/134833-why-the-3d-printing-revolution-wont-happ.en-in-your-garage) Accessed 22 August.

Widmer, R., Oswald-Krapf, H., Sinha-Khetriwal, D., Schnellmann, M. & Böni, H. (2005) 'Global Perspectives on E-Waste', *Environmental Impact Assessment Review*, 25 (5) pp. 436–58.

Wilcox, D. (2012) 'RepRap and the 3D4D Challenge' (http://youtu.be/tPvZT_OFzgo) Accessed 2 February.

Wittbrodt, B. T., Glover, A. G., Laureto, J., Anzalone, G. C., Opp.liger, D., Irwin, J. L. & Pearce, J. M. (2013) 'Life-Cycle Economic Analysis of Distributed Manufacturing with Open-Source 3-D Printers', *Mechatronics*, 23 (6) pp. 713–26.

Woudhuysen, J. (2013) 'Drilling into 3D Printing: Gimmick, Revolution or Spooks' Nightmare?' (https://http://www.dora.dmu.ac.uk/xmlui/handle/2086/8696) Accessed 19 March.

DOI: 10.1057/9781137365668.0012

Index

DOI: 10.1057/9781137365668.0013

GPSR Compliance
The European Union's (EU) General Product Safety Regulation (GPSR) is a set of rules that requires consumer products to be safe and our obligations to ensure this.

If you have any concerns about our products, you can contact us on

ProductSafety@springernature.com

In case Publisher is established outside the EU, the EU authorized representative is:

Springer Nature Customer Service Center GmbH
Europaplatz 3
69115 Heidelberg, Germany

www.ingramcontent.com/pod-product-compliance
Lightning Source LLC
Chambersburg PA
CBHW020806081025
33730CB00002B/3

9781137365651